Developing Scientific Computing Software

Jin Tang
Spence Smith

Developing Scientific Computing Software

Current Processes and Future Directions

VDM Verlag Dr. Müller

Impressum/Imprint (nur für Deutschland/ only for Germany)

Bibliografische Information der Deutschen Nationalbibliothek: Die Deutsche Nationalbibliothek verzeichnet diese Publikation in der Deutschen Nationalbibliografie; detaillierte bibliografische Daten sind im Internet über http://dnb.d-nb.de abrufbar.
Alle in diesem Buch genannten Marken und Produktnamen unterliegen warenzeichen-, marken- oder patentrechtlichem Schutz bzw. sind Warenzeichen oder eingetragene Warenzeichen der jeweiligen Inhaber. Die Wiedergabe von Marken, Produktnamen, Gebrauchsnamen, Handelsnamen, Warenbezeichnungen u.s.w. in diesem Werk berechtigt auch ohne besondere Kennzeichnung nicht zu der Annahme, dass solche Namen im Sinne der Warenzeichen- und Markenschutzgesetzgebung als frei zu betrachten wären und daher von jedermann benutzt werden dürften.

Coverbild: www.purestockx.com

Verlag: VDM Verlag Dr. Müller Aktiengesellschaft & Co. KG
Dudweiler Landstr. 99, 66123 Saarbrücken, Deutschland
Telefon +49 681 9100-698, Telefax +49 681 9100-988, Email: info@vdm-verlag.de
Zugl.: Hamilton, McMaster University, Diss., 2008

Herstellung in Deutschland:
Schaltungsdienst Lange o.H.G., Berlin
Books on Demand GmbH, Norderstedt
Reha GmbH, Saarbrücken
Amazon Distribution GmbH, Leipzig
ISBN: 978-3-639-13844-3

Imprint (only for USA, GB)

Bibliographic information published by the Deutsche Nationalbibliothek: The Deutsche Nationalbibliothek lists this publication in the Deutsche Nationalbibliografie; detailed bibliographic data are available in the Internet at http://dnb.d-nb.de.
Any brand names and product names mentioned in this book are subject to trademark, brand or patent protection and are trademarks or registered trademarks of their respective holders. The use of brand names, product names, common names, trade names, product descriptions etc. even without a particular marking in this works is in no way to be construed to mean that such names may be regarded as unrestricted in respect of trademark and brand protection legislation and could thus be used by anyone.

Cover image: www.purestockx.com

Publisher:
VDM Verlag Dr. Müller Aktiengesellschaft & Co. KG
Dudweiler Landstr. 99, 66123 Saarbrücken, Germany
Phone +49 681 9100-698, Fax +49 681 9100-988, Email: info@vdm-publishing.com
Hamilton, McMaster University, Diss., 2008

Printed in the U.S.A.
Printed in the U.K. by (see last page)
ISBN: 978-3-639-13844-3

Contents

List of Figures

List of Tables

Chapter 1

Introduction

Scientific computing (SC) software is central to our computerized society. It is used, for example, to design airplanes and bridges, to operate manufacturing lines, to control power plants and refineries, to analyze financial derivatives, to determine genomes, and to provide the understanding necessary for the treatment of cancer. Because of the high stakes involved, the quality of SC software is very important. It is essential that this kind of software has the qualities of correctness, reliability and robustness. Moreover, the complexity and size of SC problems makes the quality of performance critical. To continue to make progress in SC, and to keep the cost of software development low, SC code must also have the following qualities: usability, maintainability, portability and reusability.

To improve the quality of SC software, software engineering (SE) methodologies may be adopted. Since the late 1960s, many SE methodologies have been developed to improve software quality. These methodologies form the framework that tells us how we should develop software systems. For example, methodologies define the different phases of the development process, such as planning, requirements analysis, design, testing and maintenance. It is known that often the quality of a software system is highly influenced by the quality of the process used to acquire, develop, and maintain it; therefore, to improve the quality of software, multiple efforts have been made to improve the software development process. For instance, in the early 1980s, Watts S. Humphrey, founder of the Software Process Program of the Software Engineering Institute (SEI) at Carnegie Mellon University, created the Personal Software Process (PSP) and the Team Software Process (TSP). The goal of these processes is to improve quality and productivity in software development and to ease what was then called the "Software Crisis." His work later generated the Capability Maturity Model (CMM) and CMMI, which enables the assessment of software development processes to help improve the quality of software. Now these SE methodologies are widely accepted for developing high-quality software Wikipedia (2008).

At this time, it is unclear which of the currently available SE methodologies are most appropriate for adaptation to SC. To answer this question, it is necessary to answer

1

other questions, such as the following:

- What SE methodologies are currently used in SC?

- What technologies are currently used by the SC community?

- What qualities of SC software are in most need of improvement?

- How receptive will the SC community be to new ideas from SE?

The goal of this study is to provide insights and answers to these questions.

This chapter provides introductory information about the whole study. Section 1.1 addresses the current issues in the quality of SC software and discusses the question of why SE methodologies, widely used in business applications to improve the qualities of software, are not commonly used in SC software. Section 1.2 continues the discussion by explaining why SC software is different from other types of software. Section 1.3 clarifies the research purpose and scope. After this, the research methods that are used in the study are listed in Section 1.4.

1.1 Improving the Quality of SC Software

The quality of SC applications is very important, but the current development of SC software shows room for improvement. The field of SC has developed an impressive variety of fast and accurate algorithms and libraries. However, software qualities other than speed and accuracy have sometimes been neglected in SC code. For instance, software qualities such as usability, maintainability, portability and reusability have not always been emphasized when developing SC applications. As Dubois (2002) observes, component reuse is poor in SC, even when good mathematical libraries are available, because programmers often refuse to believe that the implementation needs to be as complicated as it is in the existing code. He also points out that not using the reliable components reduces the reliability of the software. Kreyman and Parnas (2002) argues that incomplete and imprecise documentation can potentially lead to incorrect SC software. Moreover, Wilson (2006) observes the problem that software development tools, for example symbolic debuggers, version-control systems, and systematic testers, are seldom used in developing SC software. The lack of tool use can cause problems in the development process and in the quality of the SC software.

SE has evolved steadily from its founding days until today. Now there are several systematic approaches that have been successful applied to improve software qualities. SE methodologies are now often employed in developing business applications, information systems and real-time safety critical systems.

The success of SE methodologies in other domains motivates promoting their use to improve the quality of SC. Some research has been conducted to use SE methodologies in SC software development. For instance, Dubois (2002) discusses how precisely documenting the requirements can improve the qualities of usability and maintainability of SC software. Smith (2006) defines a requirement template applied to general purpose SC software to improve its reliability, usability, verifiability, maintainability and reusability. Parker (2007) introduces developing component-based SC software and Blilie (2002) presents using patterns in scientific software to improve the reusability of the software. Unfortunately, there are only a limited number of such studies and the above methodologies have only recently been proposed and have not gained wide-spread acceptance. Moreover, it is unclear at this time which of the proposed approaches is best suited to improving the quality of SC software. A successful methodology must not only meet the technical challenges, it must also meet the social challenge of being accepted by the members of the SC community.

The question then is why is it that SE methodologies are not commonly employed in SC? and why do developers of SC software tend not to borrow ideas from SE?

One reason might be that scientists do not want to spend time on software issues that do not directly and visibly contribute to their doing science. For instance, Kelly and Sanders (2008) mentions that many SC software practitioners actually are academic scientists who develop SC software for their research purpose and scientists are primarily interested in doing science, not software. They might think that without SE methodologies, they can still develop successful SC software. Segal (2008) provides an example of a library of components for instruments embedded on a satellite. The software is developed by informal specification, i.e. face to face communication by scientists instead of formal specification that software engineers might suggest. This software was delivered on time. However, it is not yet known whether the software will perform as expected, since the instrument will not reach its destination for at least another 5 years.

Another reason that SE methodologies are not commonly employed in SC might be that given the science and engineering background of SC software practitioners, they may not know or they may not be aware of SE methodologies; therefore, they may have no sense about how SE methodologies may help them. Wilson (2006) mentions that some SC developers have simply never been shown how to program efficiently. They still use ancient text editors like Vi and Notepad to develop software and have no ideas about modern software development tools. He also mentions that, in 1998, Brent Gorda (now at Lawrence Livermore National Laboratory) started trying to address this issue by teaching a short course on software-development skills to scientists at Los Alamos National Laboratory to show them "the 10 percent of modern SE that would handle 90 percent of their needs".

Besides the above two reasons, another reason for hesitance in adapting SE method-

3

ologies might be that most SE methods and techniques do not directly map to SC applications, since the characteristics of SC software differ from those of the business and real-time systems that SE research have tendered to focus on (Smith et al., 2005). This question will be addressed further in the next section.

1.2 Overview of SC Software

If the advantages of SE methodologies were to convince SC practitioners to use the methodologies, there is still a challenge to face, since, as mentioned in previous section, SC software has characteristics that are distinct from other kinds of software. In this section, we will present an overview of SC software. First the definitions of the categories of SC software are provided, then the development process for SC software is summarized.

1.2.1 Definition of SC

For the current work, SC is defined as "the use of computer tools to analyze or simulate continuous mathematical models of real world systems of engineering or scientific importance so that we can better understand and predict the system's behavior." (Smith, 2006)

1.2.2 SC Software Categories

SC software can often be categorized as one of the following: physical model simulation software, multipurpose tools or SC environments.

Physical model simulation is the representation and emulation of a physical system or process using a computer. It can help to handle the situations that may be difficult or impossible to investigate by theoretical, observational, or experimental means. Moreover, in a wider variety of "normal" scenarios, using physical model simulation software instead of traditional "build-and-test" methods in engineering design could save time and money and decrease danger. Sometimes, physical model simulation is termed virtual prototyping (Heath, 2003, page 2).

SC software can consist of multipurpose tools, for example mathematical libraries such as NAG, SLATEC, NETLIB and IMSL. These libraries contain subroutines that are meant to be called by user-written programs. Libraries are usually written in a conventional programming language such as FORTRAN or C. These libraries provide tools such as ODE solvers, linear solvers and mesh generators. These multipurpose tools are used by the previously mentioned physical model simulations as part of the numerical solution process. The design and analysis of the mathematical algorithms on which the multipurpose tools are based are often called numerical analysis.

4

"An increasingly popular alternative for scientific computing is interactive environments that provide powerful, conveniently accessible, built-in mathematical capabilities, often combined with sophisticated graphics and a very high-level programming language designed for rapid prototyping of new algorithms" (Heath, 2003, page 35). Interactive environments like MATLAB and Maple have succeeded by integrating graphical, numerical and symbolic subsystems with a high-level problem specification language to provide rich environments for routine mathematical problem solving. These ideas have lead to a new concept in software reuse, the Problem Solving Environment (PSE). A PSE is a computer system that provides all the computational facilities necessary, such as advanced solution methods, automatic or semi-automatic selection of solution methods, and ways to easily incorporate novel solution methods, to solve a target class of problems efficiently. PSEs also include facilities to check the formulation of the problem posed, to automatically select computing devices, to view or assess the correctness of solutions, and to manage the overall computational process. Using PSEs, users can solve their problems without specialized knowledge of underlying computer hardware, software or algorithms (Rice and Boisvert, 1996). There are many PSE examples that are designed for solving SC problems. Catlin (2008) lists many such examples.

In this study, we focus on the first two categories of SC software mentioned above; that is, our focus is on physical model simulation software and multipurpose tools. Interactive environments are outside the research scope of this study.

1.2.3 Development Process for SC Software

The development processes are presented in this section for physical model simulation software and multipurpose tools. Figure 1.1 (Lai, 2004, page 3) illustrates the typical work flow for the development of physical model simulation software. The overall problem solving process in physical model simulation usually includes the following steps (Einarsson, 2005, page 13 - 15):

- From real world to mathematical model
  A mathematical model of a physical phenomenon or system is developed through doing the basic theoretical research and using assumptions to simplify the real world. It requires specific knowledge of the particular scientific or engineering disciplines involved and knowledge of applied mathematics as well. This work usually is done by domain scientists. The correctness of the mathematical model should be validated against the original problem.

- From mathematical model to computational model
  The computational model (numerical algorithm) is developed according to the mathematical model. This work is usually done by computational scientists. The

algorithms are tested against the mathematical models.

- From computational model to computer implementation
  The computational model is implemented in computer software. After this, the
  software is run to simulate the physical process numerically. This work may be
  done by computer scientists. The code should be validated against the mathematical model to ensure that the implementation is a correct reflection of the model.
  Experiments should also be conducted to validate whether the model adequately
  captures the real world phenomenon.

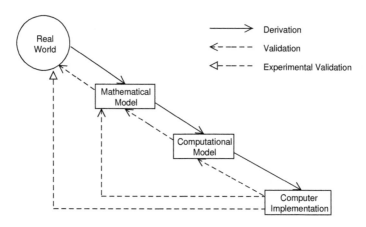

Figure 1.1: Work Flow for the Development of Physical Model Simulation Software

For multipurpose tools, the above work flow is only slightly modified, as represented by the similar work flow shown in Figure 1.2. It starts from the mathematical
model instead of the real world problem.

1.2.4 Characteristics of SC Software

The distinguishing characteristics of SC software, as compared to other types of software,
are as follows:

- Continuous Quantities
  SC deals with quantities that are continuous, as opposed to discrete. SC is concerned with functions and equations whose underlying variables - time, distance,
  velocity, temperature, pressure, and the like - are continuous in nature (Heath,
  2003, page 1).

6

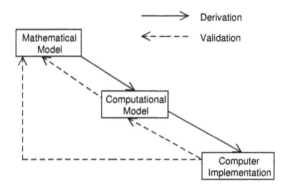

Figure 1.2: Work Flow for the Development of Scientific Multipurpose Tools

- Finite Precision

  Precision refers to the number of digits used for arithmetic, input and output. One type of approximation that is very frequently made in SC is the representation of real numbers on a computer. In a digital computer, the real number system $\mathbb{R}$ of mathematics is represented by a floating-point number system $\mathbb{F}$. The IEEE standard single-precision (SP) and double-precision (DP) binary floating-point systems are by far the most commonly used today.

- Approximation

  There are many sources of approximation or inexactness in SC. For instance, some physical features of the problem or system under study may be simplified or omitted, like friction or air resistance. Other sources of errors are properties measured by finite precision laboratory instruments and inexact input data produced by any previous computation. For the above situations, the approximations are usually beyond our control. In SC, approximation will usually be focused on truncation and rounding errors. Truncation means some features of a mathematical model are omitted or simplified. Rounding occurs when the representation of real numbers and arithmetic operations upon them is limited to some finite amount of precision, thus the solution is generally inexact. The accuracy of the final results of a computation may reflect a combination of any or all of these approximations (Heath, 2003, page 4).

- Unknown Solution

  The solutions for most SC problems are unknown. Most SC software is built to solve problems that are difficult or impossible to solve without the software. Therefore,

7

the software is the only way that a solution to the problem can be achieved. This makes judging the correctness of SC software more difficult than for many other kinds of software.

1.3 Research Purpose and Scope

According to quality issues that exist in SC software, and inspired by the success that SE methodologies have found for other kinds of software, we decided to determine what SE methodologies can be adapted for use in developing SC software, so as to improve its quality. As discussed previously, Wilson (2006) claims that only 10 percent of modern SE can be used to handle 90 percent of the needs of SC software practitioners. Although the fact is intuitively appealing and pragmatic, it would be nice to back it up with some empirical data. Our hope is that an incremental approach can be proposed for successfully introducing and promoting SE methodologies with SC practitioners. To adopt SE methodologies to SC software, the four questions listed at the beginning of this chapter should be answered. Using the answers to the above questions, we will explore what SE methodologies can be used in SC software development and how to use them to improve the quality of SC software.

1.4 Research Methods

This section describes the research methods that were applied to solve the questions that have been previously mentioned. Concretely, the methods associated with their corresponding questions are as follows:

1. What SE methodologies have been specifically proposed for SC?

   A literature search was used to obtain information about the currently available SE methodologies that might be used in SC software. The overview of SE methodologies is presented in Chapter 2.

2. What SE methodologies are currently used in SC? What technologies are currently used in SC communities? What qualities of SC software are in most need of improvement? How receptive will the SC community be to new ideas from SE?

   To solve the above questions, a survey titled "Survey on Developing Scientific Computing Software" was conducted. This survey used a literature search and qualitative and quantitative research methods. An open source software package called Surveypro (eSurveysPro, 2008) was used to conduct this online survey. There are four phases in this survey, which are questionnaire design, survey pilot testing, an online survey and survey data analysis. This survey was approved by the McMaster University Research Ethics Board. Details of the above four phases is presented in Chapter 3 and 4, respectively.

3. How to use SE methodologies to improve the quality of SE software?

As mentioned in Section 1.1, the disconnect between SE research and SC applications is widened by the fact that there are so few examples available in the literature that illustrate how the two can be combined. This study attempts to address this problem by showing how SE methodologies can be adapted to SC applications using a one-dimensional Numerical Integration Solver (ONIS) as an example. ONIS is developed following two different software developing processes: i) Modified Parnas' Rational Design Process, and ii) Unified Software Development Process. The advantages and disadvantages of these two processes are compared in Chapter 5, to help convince industry and academia to use SE methodologies to develop SC software. Moreover, the comparison provides insight into a strategy for incremental adaptation of SE methodologies.

Chapter 2

Software Engineering Methodologies

The goal of this research is to improve the quality of SC software through the use of SE methodologies. To understand the current SE options available, this chapter reviews the literature on SE methodologies. In addition, a discussion is included on the few SC specific SE methodologies that have been proposed. The discussion focuses on how well these proposed methodologies might fit with the special characteristics of SC software, as presented in the previous chapter.

As mentioned in Chapter 1, a good development process will often help to improve the quality of the software. However, a systematic software development process is rarely used in SC software development. Guatelli et al. (2005) mentions SC software development teams that did not follow any software process explicitly, which left the achievement of the project goals entirely to the personal efforts of the individual developers. Further, the project suffered from repeated failure to match the release schedule, and the released versions were often unstable when used in analysis applications. One reason for a lack of a systematic process may be that many SC programs are developed as in-house software. Developers tend to create software for their own usage, or for use in the experiments they are involved in, rather than software products to be deployed to customer companies or to the general public. Given the limited audience, and potentially limited lifetime of the software, a systematic process including documentation may not be proposed or adhered to. Another more important reason for a lack of a systematic process might be, as addressed in Chapter 1, that the SC software developers are usually not software professionals, so they do not have a formal education on the software development process.

The process models currently popular in SE are summarized at the beginning of this chapter. Unfortunately, these options do not seem to be popular in SC, as in most cases, when a systematic software process is adopted, it does not follow any of the widely

11

known models (Guatelli et al., 2005). The reason for this might be that the existing software process models were not designed especially for SC problems. If a software process is provided according to the characteristics of SC problem, this might help the SC community adopt it.

There are five sections in this chapter. Section 2.1 presents the development process models of the waterfall model and the evolutionary development. Section 2.2 analyzes Modified Parnas' Rational Design Process (PRDP) and Section 2.3 discusses the Unified Software Development Process (USDP). Section 2.4 illustrates some common SE methods for software reuse, such as component-based SE, mathematical software libraries and design pattern. Finally, Section 2.5 provides software tools that might be helpful when developing SC software.

2.1 Software Development Process Model

Software development is a creative and a step-by-step process, often involving many people producing many different kind of products. The software development process is sometimes called the software life cycle, which usually involves the following stages (Pfleeger and Atlee, 2006): requirements analysis and definition, system design, program design, program implementation, unit testing, integration testing, system delivery, and maintenance.

A software process model is an abstract representation of a software process that presents one view of that process. Each process model represents a process from a particular perspective, and thus provides only part of the information about that process (Sommerville, 2004, page 65). Process models may include activities that are part of the software process, software products and the roles of people involved in software engineering. Building process models and discussing the associated subprocesses helps the team to understand the gap between what it should be and what it is. Most software development process models start with system requirements as input and all processes end with delivered products as output.

Here, we only present a number of very general process models from an architectural perspective. That means we provide the framework of the process, but not the details of specific activities. The process models introduced here are as follows: the Waterfall Model, Evolutionary Development and Component-Base Software Engineering. This section focuses on the first two models, especially the Waterfall Model, since PRDP and USDP, which will be introduced in Sections 2.2 and 2.3, are based on this model. Component-Based Software Engineering will be introduced later in Section 2.4 on Software Reuse. These three generic process models are widely used in current software engineering practice. They are not mutually exclusive and are often used together, especially for the development of large systems.

2.1.1 The Waterfall Model

The waterfall model was proposed by Royce (1970). Figure 2.1 illustrates this model. The principal stages of the model map onto five fundamental development activities: *requirements analysis and definition, system and software design, implementation and unit testing, integration and system testing* and *operation and maintenance*.

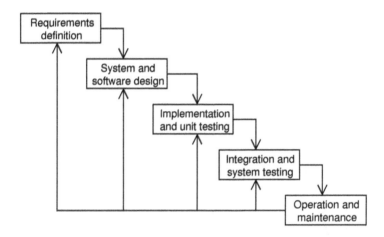

Figure 2.1: The Software Life Cycle

In principle, the result of each phase is one or more documents that need to be approved. The following phase should not start until the previous phase has finished. In practice, these stages overlap and provide information to each other. For example, during design, problems with requirements are identified; during coding, design problems are found and so on. Therefore, the software process is not a simple linear model but involves a sequence of iterations of the development activities.

The advantages of the waterfall model are that documentation is produced at each phase. Its major problem is its inflexible partitions, which divide the project into distinct stages. Commitments must be made at an early stage in the process, which makes it difficult to respond to changing customer requirements. Therefore, the waterfall model should only be used when the requirements are well understood and are unlikely to change radically during system development (Sommerville, 2004, page 67).

The waterfall model turns out to be well suited to SC problems. The reasons are as follows:

- As mention above, the waterfall model works well when the requirements are stable, which is certainly the case for the SC problems, since scientific theories are slow to change and numerical analysis methods are mostly well established.

- From Figure 1.1 and 1.2, we find that the typical work flows for development of physical model simulation software and multipurpose tools also follow a waterfall process.

- As presented above, documentation plays an important role in the waterfall model and this characteristics makes it fit SC problems well. Kreyman and Parnas (2002) mention that because of the complexity of mathematical models that are usually used in engineering and scientific simulations, good documentation, which is easy to be understood and used during software development, could dramatically reduce the number of errors in the software.

Using the waterfall model can improve the quality of software, because traceability is included between the stages of the waterfall model and traceability helps to improve the verifiability of completeness and consistency, which in turn improves the reliability of the software. Moreover, traceability helps to manage proper changes in the software, since the connection between design and the anticipated changes is also provided in the documentation. Not surprisingly, traceability can improve the maintainability of SC software.

2.1.2 Evolutionary Development

Evolutionary development is based on the idea of developing an initial implementation, exposing this to user comment and refining it through many versions until an adequate system has been developed. There are two fundamental types of evolutionary development (Sommerville, 2004, page 68):

1. Exploratory development
   The development of the software starts with the parts of the system that are understood. The system evolves by adding new features proposed by the customer.

2. Throwaway prototyping
   The purpose of the evolutionary development process is to understand the customer's requirements and hence to develop a better requirements definition for the system. The prototype focuses on the experiments where the customer's requirements are poorly understood.

Compared with the waterfall approach, the evolutionary approach is more effective in meeting the immediate needs of the customers. Also, it facilitates developing specifications incrementally. However, the evolutionary approach has two problems. One is that it causes the potential problem of missing documentation. If consecutive versions of a system are developed quickly, it is not cost-effective to produce documents that reflect every version of the system. The other problem is that sometimes the evolutionary approach leads to a systems with poor structure, because continual change tends to corrupt the software structure. Moreover, incorporating software changes increases the difficulties and cost of the software development.

The evolutionary approach can also be used to develop SC software, especially for small and medium-sized system. Actually the evolutionary approach is the way that many SC programs are developed (Morris, 2008). For example, when people develop a two or three dimension partial differential equation solver, they might start from the most understandable part to create a one dimensional (1D) model first. After the 1D model, new features are added to make the software use a 2D or 3D model. For large systems, a mixed process, incorporating the best features of the waterfall and the evolutionary development model would be a better choice. Developers may develop a throwaway prototype using an evolutionary approach to resolve uncertainties in the requirement specification. After the requirement is well specified, the waterfall based process can be used, so that the software can grow in a predictable and manageable manner.

2.2 Modified Parnas' Rational Design Process (PRDP)

In 1986, Dr. Parnas developed a rational design process (Parnas and Clements, 1986), which divided the software development process into seven steps: 1) establish and document requirements; 2) design and document the module structure; 3) design and document the module interfaces, 4) design and document the uses hierarchy, 5) design and document the module internal structures, 6) write programs and 7) maintain. PRDP focuses on the documents in each stage. In particular, it uses formal specification to document the requirements to avoid ambiguous requirements. A relational requirements model and program document model based on this rational design process has been successfully applied to the safety shutdown systems of the Darlington Nuclear Power Generating Station in Ontario (Parnas et al., 1991).

Some efforts have been made to improve PRDP for better use in SC software development. For example, in PRDP, although Dr. Parnas addresses the requirement specification and also provides ideas about how to write a requirement specification, he did not provide a template that developers can follow to create their own requirements document for SC. To address this omission, Smith et al. (2005) develop a new requirements template for scientific computing software and Smith (2006) modifies the previous template to provide a new template for multi-purpose tools for SC software. Moreover, Yu (2007) provides an example of using a modified PRDP to develop a parallel mesh genera-

15

tion toolbox. The modified PRDP includes six stages, which are 1) commonality analysis; 2) software requirement specification; 3) module guide; 4) module interface specification; 5) implementation and 6) testing. We will refer to this modified PRDP as simply PRDP throughout the rest of this book.

To help the reader better understand PRDP, basic knowledge of the main procedures are introduced in this section. In addition, an example of a one dimension numeric integration solver (ONIS), is presented in this book. Appendix B presents the documentation of ONIS using PRDP.

2.2.1 Commonality Analysis (CA)

Commonality Analysis (CA) is a process to study shared features or attributes among similar software products to find possibilities for development of the software as a program family. This analysis is performed before the software requirements activities. Using CA can help to reuse the common aspects of family members so as to rapidly develop new family members. The idea of program families initially came from Dijkstra (1972). To apply CA in SC software development, a CA template is provided in Smith (2006).

2.2.2 Software Requirement Specification

Software requirements activities usually include: i) software requirement analysis, a process to understand what the software is supposed to do as well as to refine and model the requirements; ii) software requirement documentation, which involves writing a document containing a complete description of what the software will do, without describing how it will do it (Davis, 1990); iii) software requirement verification, a stage to check whether the requirements are consistent and complete.

In PRDP for SC, the software requirement specification (SRS) follows a template provided in Smith (2006), which is designed especially for SC problems. There are 11 sections in this template, which are: 1) Reference Material; 2) Introduction; 3) General System Description; 4) Specific System Description; 5) Non-functional Requirement; 6) Solution Validation Strategies; 7) Other System Issues; 8) Traceability Matrix; 9) List of Possible Changes in the Requirements; 10) Values of Auxiliary Constants; 11) References. The characteristics and advantages of this template can be found in Smith (2006) and Smith et al. (2005).

This template is designed for SC software. For example, some specific sections in this template, such as the theoretical model, sensitivity of the model and tolerance of the solution, can help present SC problems. This template can be widely used to document different kinds of SC software. For example, Lai (2004) uses it to document the SRS for the specific case of engineering mechanics software and Yu (2007) applies this template to document a scientific library for mesh generation. Moreover, an additional example, SRS for a one dimension integration solver, is provided in this book in Appendix B.1.

2.2.3 Module Guide (MG)

In PRDP, in the design stage, the system is decomposed into modules according to the *information hiding* principle. The independence of modules will improve the reusability of the software. Moreover, information hiding helps designers and maintainers focus on their working parts and not worry about the irrelevant parts, in this way, the maintainability and flexibility of the software are improved.

This stage, which is named as MG in PRDP, belongs to software architectural design, i.e. analyzing the overall structure of the software and the ways in which that structure provides conceptual integrity for a system (Shaw and Garlan, 1995, Page 307). The MG document can include the following sections: 1) Connection Between Requirements and Design; 2) Anticipated Changes; 3) Unlikely Changes; 4) Module Hierarchy; 5) Module Decomposition; 6) Traceability Matrix; and 7) Use Hierarchy between Modules. Appendix B.2 presents the MG for ONIS, as well as an introduction to this template.

2.2.4 Module Interface Specification (MIS)

Module interface design belongs to detailed design. In this stage, a document named the MIS is produced. The MIS describes what the module will do, but not how to do it. Hoffman and Strooper (1999) presents a template to document MIS. Yu (2007) provides an example for a mesh generator that makes some modifications to this template. In the MIS, each module can be regarded as a finite state machine which has a set of state variables, inputs, outputs, and transitions. There are four sections in this template, which are *Module name*, *Uses*, *Interface Syntax* and *Interface Semantics*. *Module Name* gives the name of the module. *Uses* lists constants, data types and access program that are defined outside of this module. *Interface Syntax* defines the syntax of the module interface. *Interface Semantics* introduces the semantics associated with the above syntax. The above is a very brief introduction of this template, detailed information can be found in (Yu, 2007, Page 50 - 60). Appendix B.3 presents the MIS for ONIS.

2.3 Unified Software Development Process (USDP)

The USDP is one of the most widely adopted process frameworks in the industrial environment. It is use-case driven, architecture-centric, iterative and incremental. The USDP uses the Unified Modeling Language (UML) when preparing all blueprints of the software system. A formal constraint language, called the Object Constraint Language (OCL) is defined in UML to describe expressions on UML models. The characteristics of USDP, UML and OCL are presented later in this section.

Currently, USDP is increasingly used in SC software. The following are some successful projects where USDP has been adopted: Geant4 Low Energy Electromagnetic Physics (Chauvie, 2004), which is a simulation software project; Statistical Toolkit (Cirrone et al., 2004), which is a mathematical library for data analysis; Brachytherapy Dosimetry System (Guatelli, 2004), which is simulation for oncological radiotherapy; and, Bepi Colombo simulation (Mantero, 2004). However, from a literature search, it is hard to find a complete example that people can follow to develop their own SC software using USDP. In this book, we provide an example that uses USDP to develop and document ONIS. Appendix C provides the documents for this process.

Using USDP can help successful delivery of the software. Guatelli et al. (2005) mentions that if we defined a successful project as one completed on time, delivering all the planned features and functions, respecting the original budget, the success rate measured on the physics research projects adopting the USDP is 95%, comparing to the success rate of 23% reported by the Standish Group. This may explain why USDP is becoming more and more popular in current software development practice.

2.3.1 Characteristics of USDP

The real distinguishing aspects of the USDP are captured in the three key words: use-case driven, architecture-centric, and iterative and incremental (Jacobson et al., 1999, Page 3 - 8).

- The USDP is Use Case Driven

  In USDP, the term *user* refers not only to human users but to other systems. A use case is a piece of functionality in the system that gives a user a result of value. Use cases capture functional requirements. All the use cases together make up the use-case model, which describes the complete functionality of the system and drives the system's design, implementation and testing, in other words, use cases drive the development process.

- The USDP is Architecture-Centric

  Software architecture is "the set of significant decisions about the organization of a software system, the selection of the structural elements and their interfaces by which the system is composed, together with their behavior, as specified in the collaborations among those elements. The architecture also includes the composition of these structural and behavioral elements into progressively larger subsystems, and the architecture style that guides this organization, which includes these elements and their interfaces, their collaborations, and their composition. Software architecture is concerned not only with structure and behavior but with usage, functionality, performance, resilience, reuse, comprehensibility, economic and technology constraints and trade-offs, and aesthetic concerns" (Jacobson et al., 1999, Page 443). Architecture-centric means that in the software life cycle, the system's architecture is used for conceptualizing, constructing, managing and evolving the system under development.

- The USDP is iterative and incremental
  When developing large software systems, it is practical to divide the work into smaller slices or mini-projects. Each mini-project is an iteration that results in an increment. Iterations refer to steps in the workflow, and increments, to growth in the product. The iterations should be controlled; that is, they should be selected and carried out in a planned way. In every iteration, the developers identify and specify the relevant use cases, create a design using the chosen architecture as a guide, implement the design in components, and verify the components to satisfy the use cases. If an iteration meets its goals, development goes to the next iteration; otherwise, the developers should revisit their previous decisions and try a new approach.

In USDP, use-case driven, architecture-centric, and iterative and incremental development are equally important. Architecture provides the structure that guides the work in the integration, and use cases define the goals that drive the work of each iteration.

2.3.2   The Unified Modeling Language (UML)

UML is a standard modeling language for visualizing, specifying, constructing, and documenting software systems. UML is informed by a vision of the structure of software systems known as the 5 view model. The 5 views are: *use case view, design view, implementation view, deployment view* and *process view*. The *use case view* defines the system's external behaviour. This view defines the requirements of the system, and therefore constrains all the other views. The *design view* describes the logical structures that support the functional requirements expressed in the use case view. The *implementation view* presents the physical components out of which the system is to be constructed. The *process view* deals with issues of concurrency within the system, and the *deployment view* describes how physical components are distributed across the physical environment (Priestley, 2003, Page 7 - 10).

In UML, the information relevant to each view is recorded in the various types of models. For example, a use case model presents the information in the use case view. Models may also be produced at different level of abstraction. A model is normally presented to a designer as a set of diagrams. Diagrams are graphical representations of collections of model elements. UML defines nine distinct diagram types, which are listed in the Table 2.1 (Priestley, 2003, Page 10) together with an indication of the views with which each is characteristically associated.

Diagram	View
Use case diagram	Use case view
Object diagram	Use case and design view
Sequence diagram	Use case and design view
Collaboration diagram	Use case and design view
Class diagram	Design view
Statechart diagram	Design and process view
Activity diagram	Design and process view
Component diagram	Implementation view
Deployment diagram	Deployment view

Table 2.1: UML's Diagram Types

In the documents for ONIS using USDP, only some of the diagrams in Table 2.1 are employed. The introduction and usage for those diagrams can be found in Appendix C.1 and C.2.

2.3.3 The Object Constraint Language (OCL)

A UML diagram is typically not refined enough to provide all of the relevant aspects of a specification; therefore, additional constraints about the objects are needed in the model. Such constraints are often described in natural language. However, practice has shown that this will always result in ambiguities. To write unambiguous constraints, a formal language, called OCL, has been developed to express constraints.

Unfortunately, it is very rare to find an example about using OCL in UML to document USDP when developing SC software. Appendix C.2 provides one example. In Appendix C.2, the system design specification (SDS) for ONIS using USDP, most of the constraints are documented using OCL, but there are still some constraints written using natural languages. The introduction and usage of OCL expressions, which are used in the SDS, are presented in Appendix C.2.

2.4 Software Reuse

Software reuse can help to lower software production and maintenance costs, speed up delivery of systems and improve reliability. The software units that are reused may be at different sizes and different levels. For example (Sommerville, 2004, Page 416):

- Application system reuse

  The whole application system may be reused by incorporating it into other systems by configuring the application for different customers or by developing program families that have a common architecture, but are tailored for specific customers.

- Component reuse

  Components are designed for the purpose of software reuse. The size of components ranges from sub-systems to single objects.

- Object and function reuse

  A mathematical function or an object class may be reused in software system.

2.4.1 Component-based Software Engineering (CBSE)

In Component-based software engineering (CBSE), units of software functionality are encapsulated as components that interact with each other through well-defined interfaces. The actual implementation is opaque to other components, and application composition is achieved through providing and using these interfaces. Component-based environments typically offer a "plug and play" approach to composition of component into applications, in which components offering the same interface are interchangeable, allowing easy swapping of components to test new algorithm, tune for performance, and other reasons (Alexeev et al., 2005). Figure 2.2 illustrates the process of CBSE.

Figure 2.2: Component-based Software Engineering

While components have been explored for many years in SE, their application for SC has only just begun to unfold. Alexeev et al. (2005) mentions, since 2001, that the SciDAC-funded Center for Component Technology for Terascale Simulation Software has led a program of research and use of component technologies in high-performance SC. Decker and Johnson (1998) introduces using component to create problem-solving environment to solve linear algebra problems. Kellner (2005) presents 3D simulation software, called ORCAN, that has developed based on component technology.

The advantage of using CBSE is that it helps improve productivity and performance. For example, the SciDAC Center for Reacting Flow Science found that, using a component-based approach, their productivity and performance has increased greatly (Lefantzi et al., 2004). Similarly, the quantum chemistry community has achieved order-of-magnitude performance improvements (Kenny et al., 2004). Moreover, CBSE can help reduce the amount of software to be developed so as to reduce cost and risks. It usually leads to faster delivery of the software. However, Figure 2.2 shows that requirements compromises are inevitable and this might lead to a system that does not meet the real needs of the users. The reason that CBSE inevitable leads to requirements change is that in the requirements specification stage, the requirements focus on the functionalities of the system instead of on reusing the system. In the next stage, *component analysis*, emphasizes is on system reuse, which includes searching existing components to match the given system and designing new components when none currently exist; Therefore, the requirement modification stage shown in Figure 2.2 is necessary, because when components are decided, requirements are updated using information about the components that have been discovered or newly designed. To reuse available component, requirements compromises are inevitable and this might lead to a new system that diverges from the original needs.

2.4.2   Mathematical Software Libraries

Mathematical Software libraries were introduced in the 1960s to support the reuse of high quality software. The increasing number, size and complexity of mathematical software libraries made possible the development of complex SC software with high efficiency. In ONIS, routines from the Quadpack library are used for integration. The followings are libraries that are often used in SC software development (NPL, 2008).

- NAG: A large FORTRAN Library covering most of the computational disciplines including quadrature, ordinary differential equations, partial differential equations, integral equations, interpolation, curve and surface fitting, optimization, linear algebra, correlation and regression analysis, analysis of variance and non-parametric statistics.

- IMSL: International Mathematical and Statistical Libraries which is similar to NAG.

- LINPACK: A FORTRAN library for solving systems of linear equations, including least-squares systems.

- EISPACK: A companion library to LINPACK for solving eigenvalue problems.

- LAPACK: This is a replacement for, and further development of, LINPACK and EISPACK. LAPACK also appears as a sub-chapter of the NAG library.

- Harwell: Optimization routines including those for large and/or sparse problems.

- DASL: Data Approximation Subroutine Library, developed for data interpolation and approximation with polynomial and spline curves and surfaces.

22

- MINPACK: A FORTRAN Library developed for function minimization. MINPACK contains software for solving nonlinear least-squares problems.

2.4.3 Design Patterns

A "Design pattern describes a problem which occurs over and over again in our environment, and then describes the core of the solution to that problem, in such a way that you can use this solution a million times over, without ever doing it the same way twice" (Alexander and Ishikawa, 1977). Gamma et al. (1995) define the four essential elements of design patterns:

- Pattern Name: the pattern name is a meaningful reference to the pattern.

- Problem: the problem is the description that describes when to apply the pattern.

- Solution: the solution is the parts of the design solution. It describes the elements that make up the design, their relationships, responsibilities and collaborations. The solution is not a concrete design or implementation; it is at an abstract level, since a pattern is a template that can be applied in many different situations.

- Consequences: the consequences are statements about the results and trade-offs when applying the pattern. The consequences of a pattern include its impact on a system's flexibility, extensibility, or portability.

Research about design patterns have been conducted in the SC community. Fowler (1997) discusses several patterns directly applicable to scientific software, which include measurements, quantities with units, observation processes and hypotheses. Cickovski et al. (2004) also presents a couple of patterns, for example particle-mesh pattern, multiple space pattern, and so on, which are used in a molecular dynamics software system. Blilie (2002) mentions how to design and apply a grid or mesh pattern in dynamic systems.

2.5 Software Tools

In the whole software development process, software tools are very useful for designing and coding the system, capturing requirements, organizing documentation, testing, keeping track of changes, and more. These tools are sometimes called Computer-Aided Software Engineering (CASE) tools. CASE tools assist SE practitioners in every activity associated with the software process. CASE can be as simple as a single tool that supports a specific SE activity, or as complex as a complete "environment" that encompasses tools, a database, people, hardware, a network, and an operating system. The followings are major categories of CASE tools (Pressman, 2001):

1. Process modeling tools: The tools are used to represent key elements of a process to help better understand the development process.

2. Project management tools: The tools help to track and monitor the project schedule and project plan.

3. Requirements tracing tools: The objective of requirements tracing tools is to help the final delivered system meet the requirements specification. So, the tools provide a systematic approach to the isolation of requirement. The typical requirements tracing tool combines human interactive text evaluation with a database management system that stores and categorizes each system requirement that is "parsed" from the original specification.

4. Documentation tools: The tools help to produce documentation of software system to improve the productivity.

5. Analysis and design tools: The tools enable people to create models of the system to be built. By performing consistency and validity checking on the models, analysis and design tools help to eliminate errors in those models.

6. Prototyping and simulation tools: The tools are used to predict the behavior of a software system prior to the time that it is built. They also help customers to gain insight into the function, operation and response prior to actual implementation.

7. Interface design and development tools: The tools are actually a tool kit of software components such as menus, buttons, window structures, icons, scrolling mechanisms, and so forth.

8. Programming tools: The tools encompass the compliers, editors, and debuggers that available to support most conventional programming languages.

9. Test management tools: The tools are used to control and coordinate software testing for each of the major testing steps.

10. Reengineering tools: The tools are used for reverse engineering, such as taking source code as input and generating graphical structured analysis and design models.

Many CASE tools, which are used in business applications, can also be used to enhance the productivity of SC software design and development. Kim et al. (2004) discusses how a couple of these tools, such as IBM Rational Rose and Microsoft Visio, which support UML modeling, could be used for SC software design. Warrier et al. (2008) lists several code development tools, debugging tools and version control tools to help SC software development. Sometimes, people will create tools by themselves. For example, Luksch et al. (1996) addresses LRR-TUM, a tool to design and analyze parallel programs.

Tools are also used in the development of ONIS. For example, an open source software package, umlet8 (Umlet, 2008), is used in the design stage to create UML models. Microsoft Visual Studio 2005 with build-in compile, edit and debug tools helps to develop and debug ONIS. To trace the change of documentation for ONIS, a version control tool, subversion is used to keep track of the different documents.

Chapter 3

Survey on Developing Scientific Computing Software

As mentioned in Chapter 1, given the important applications of SC software, it is surprising that SE methodologies are often neglected when developing SC software. The neglect of SE methodologies potentially causes problems with the quality of SC software. To improve the quality of SC software, we first need to know the current approach to developing SC software. Concretely, we need to solve the following questions, which were already mentioned in Chapter 1: 1) What SE methodologies are current used in SC? 2) What technologies are currently used in SC communities? 3) What qualities of SC software are in most need of improvement? 4) How receptive will the SC community be to new ideas from SE? To answer these questions, a survey titled "Developing Scientific Computing Software" was conducted. The survey questions focus on the SC software development process including methodologies and technologies that pople are using to develop SC software; therefore, the answers of previous two questions are explicit.

In this chapter, first, the goal of survey is presented. To clarify the goal, 31 research issues (RI) are designed. RIs are also a guide to design the survey questions; in other words, the survey questions are designed and developed following the RIs with the aim of finding answers to these RIs. RIs are presented in Section 3.1. To clarify the relationships between the RIs and the survey questions behind each RI the associated survey questions are listed in the brackets. There are 37 questions in the online survey. Most survey questions are actually the same as the RIs, but using a different wording to transform the issue into a survey question. An example of this is how, "RI1: Before development, do they set up a project schedule or project plan?" becomes Question 15 "In your group, do you set up a project schedule or a project plan before developing software? Yes, No". In some cases, from first appearances, the survey question is different from its associated RI, but actually the answer for the RI is implicit in the answer of the survey question. For instance, RI23: "Do they use tools to test? what kind of tools do they use?" has the associated survey question, Question 22, "A programming tool or software tool is a program or application that software developers use to create, debug, or maintain other programs and applications. When you develop software, where do you use tools? Select all that apply from: Never use tools, Design software, Code generation, Debug code, Documentation generation, Unit testing, Integration testing, Version control, Others." From analyzing Question 22, we can get the answer for RI23.

After identifying RIs, the type and content for each survey question are decided according to the RIs, and keeping in mind the four questions that were posed at the beginning of this chapter. The survey questions emphasize the SC software development process including SE methodologies and technologies which are currently used in developing SC software; therefore, the answers for the first two questions on page 43 are explicit. For the third question, about the qualities in most need to improvement, the answer implicitly provided by analyzing the survey results, as presented in Chapter 4. To learn how receptive the SC community to be new ideas which addresses Question 4, we designed two fill-in questions, Question 34 and 35. From the respondents' comments for these two questions, we can get some sense of how willing they are to modify their development process. In Section 3.2, survey questions are presented by an index table which provides the topics that each question emphasizes. Moreover, question types that are adopted in this survey are illustrated by examples.

Section 3.3 discusses the entire survey process which consists of four phases. They are questionnaire design, survey pilot testing, online surveying and survey data analysis. Because questionnaire design was provided in Section 3.1 and survey data analysis will be presented in Chapter 4, Section 3.3 focuses on the other two phases: survey pilot testing and online surveying.

The survey questionnaire design and survey process were inspired by the following sources: (Foreman, 1991), (Forward, 2002) and (Statistics Canada, 2005).

3.1 Goal of the Survey

The short term goal of this survey is to find the processes that industry and academia follow when developing their SC software. The mid term objective is to direct research on identifying potential shortcomings of current SC software development approaches and adopting SE methodology to improve the quality of SC software.

To solve the four questions mentioned in the introduction and to clarify the goals of the survey, before the questionnaire was designed, 31 questions, which are presented below, were posed. Many of these questions were motivated by the waterfall model. This means that we divided the whole software development process into five stages, which are requirements, design, coding, testing and maintenance. From the survey results, we can obtain an overall picture about how SC software is currently developed. In addition to RIs on the software development process, RIs were also posed on software documentation, as this is an integral part of many SE methodologies. Finally, RIs were posed about the developer's education background and working experience. An understanding of the background of practitioners is necessary when proposing any changes to the current development process.

The following are the 31 RIs posed, divided into 8 sections, which we intended to address through the survey.

Software Development Process
In a SC software development group:

RI1: Before development, do they set up a project schedule or a project plan? (Associated with Question 15)

RI2: Do they use process models? What kind of models do they use? (Associated with Question 16)

RI3: What is the time distribution of the whole software development process? (Associated with Question 23)

For different types of organizations, different size groups and different size software, what is the difference in the answers to the above questions?

Requirements
In a SC software development group,

RI4: Do they have requirements specifications? (Associated with Questions 17 and 27)

RI5: What kind of specification do they provide? (Associated with Question 17)

RI6: Do they use semi-formal specification? What kind of semi-formal specification do they use? (Associated with Question 18)

RI7: Do they use formal specification? What kind of formal specification do they use? (Associated with Question 19)

RI8: What kind of non-functional requirements are important for SC software? What is the order of importance of these non-functional requirements? (Associated with Question 13)

For different types of organizations, different size groups and different size software, what is the difference in the answers to the above questions?

Design
In a SC software development group:

RI9: Do they have design documentation? What kind of design documentation do they provide? (Associated with Question 27)

RI10: Do they consider software reuse in the design stage? What kind of software reuse methods do they use? (Associated with Question 21)

RI11: For software reusability, what kinds of libraries do they use? (Associated with Question 12) For different software types and different software application fields, what is the difference in the answers to this question? (Associated with Questions 7 and 8)

RI12: Do they use tools to help design software? What kind of tools do they use? (Associated with Questions 22 and 33)

RI13: Do they consider a testing plan in the design stage? (Associated with Question 25)

For different types of organizations, different size groups and different size software, what is the difference in the answers to the above questions?

Coding

In a SC software development group:

RI14: Do they have coding standards that the whole group needs to follow? (Associated with Question 20)

RI15: Do they use tools to generate code automatically? (Associated with Questions 22 and 34)

RI16: Do they use tools to debug programs? What kind of debug tools do they use? (Associated with Questions 22 and 34)

RI17: Do they use version control tools? What kind of version control tools do they use? (Associated with Questions 22 and 34)

RI18: What kind of source code languages that people are using to develop SC software (Associated with Question 9)

RI19: What kind of operating systems are people using to develop SC software (Associated with Question 10)

For different types of organizations, different size groups and different size software, what is the difference in the answers to the above questions?

Testing

In a SC software development group:

RI20: How do they choose test cases? (Associated with Question 25)

RI21: What kind of validation and verification methods do they use? (Associated with Question 24)

RI22: Do they generate testing reports? (Associated with Question 27)

RI23: Do they use tools to test? What kind of tools do they use? (Associated with Question 22 and Question 34)

RI24: Are there specific people in charge of testing? (Associated with Question 26)

For different types of organizations, different size groups and different size software, what is the difference in the answers to the above questions?

Maintenance

In a SC software development group:

RI25: What is the life time of typical SC software? (Associated with Question 14)

Documentation

In a SC software development group:

RI26: Do they use tools to generate documentation? What kind of tools do they use? (Associated with Question 22 and Question 34)

RI27: How often do they update documentation? (Associated with Question 29)

RI28: How is good documentation built? (Associated with Question 30)

RI29: What are factors causing documentation to be out of sync with the system it describes? (Associated with Question 30)

For different types of organizations, different size groups and different size software, what is the difference in the answers to the above questions?

People

RI30: What kind of education background is needed when developing SC software? (Associated with Question 3)

RI31: What kind of working experiences is needed for people who develop SC software? (Associated with Questions 5 and 6)

For different types of organizations and different size groups, what is the difference in the answers to these questions?

3.2 Questionnaire Design

To achieve the research goals and solve the above RIs, a questionnaire was designed. The questionnaire has 37 individual questions, as shown in Appendix A.1. This section summarizes the design of the questions and their format.

3.2.1 Question Design

The 37 questions in the questionnaire are divided into six sections, which are *characterization of participant*, *characterization of the SC software that the participant is typically involved with developing*, *methodology*, *testing*, *documentation* and *feedback*. Table 3.1 and Table 3.2 present the question topics and their associated question numbers for each section in the questionnaire.

3.2.2 Question Types

There are four types of questions in this survey, which are *multiple choice multiple answer question*, *multiple choice single answer questions*, *rating question* and *fill-in question*. The

Section	Question No.	Content
Characterization of Participant	1	Organization type
	2	Group size
	3	Education background
	4	Job functions of a group
	5	Working experience (working years)
	6	Working experience in programming
Characterization of the SC software that the participant is typically involved with developing	7	Software application fields
	8	Software type
	9	Source code languages
	10	Operation systems
	11	Software sizes
	12	Mathematical Libraries used in developing software
	13	Non-function requirements
	14	Software Lifetime

Table 3.1: Questions in the Questionnaire 1

first three types of questions are closed questions (ChangingMinds, 2008), which means users can answer this kind of questions with either a single word or a short phrase. A fill-in question is an open question, which is used to receive a long answer to get information about respondents' opinions and feelings. *Multiple choice multiple answer question* and *multiple choice single answer questions* are very similar. The distinction between them is that the first one allows the participant to select more than one answer. *Rating question* is intended to help scale the answers. The followings are samples for each of the four questions types.

- Multiple Choice Multiple Answer Question (more than one selection is allowed)

    What types of SC software are you involved in developing? Please select all that apply.
    ☐ Fast Fourier Transform
    ☐ Interpolation
    ☐ Linear Solver
    ☐ Linear Least Squares
    ☐ Mesh Generation
    ☐ Numerical Integration
    ☐ Optimization
    ☐ Ordinary Differential Equations (ODE) Solver
    ☐ Random Number Generator
    ☐ Partial Differential Equations (PDE) Solver
    ☐ Stochastic Simulation

Section	Question No.	Content
	15	Project plan
	16	Process model
	17	Type of specifications
	18	Semi-formal specification
Methodology	19	Formal specification
	20	Coding standard
	21	Software reuse
	22	Tools
	23	Development process
	24	Validation and verification methods
Testing	25	Test cases
	26	Who is in charge of the testing phase
	27	Type of documentation
Documentation	28	Speed with which documentation is updated
	29	Factors of good documentation
	30	Factors causing documentation to be out of sync with the system it describes
	31	Interest in receiving this survey
Feedback	32	Interest in receiving a phone interview
	33	Personal contact information
	34	Specifying tools
	35	Addition comments on software qualities, software documentation
	36	Specifying process improvement
	37	Remark

Table 3.2: Questions in the Questionnaire 2

☐ Solving Eigenvalues

☐ Solving Nonlinear Equations

☐ Others (identify):

- Multiple Choice Single Answer Question (one and only one selection is allowed)
  How many people in your current group are involved in developing SC software?
  ○ 1
  ○ 2 - 5
  ○ 6 - 15
  ○ 16 - 50
  ○ 51 - 100
  ○ > 100

- Rating Question
  In your experience, how important is each of the following software qualities. Please

rate the relative importance of the qualities, with one (1) for the LEAST important items and five (5) for the MOST important. If you feel that there are software qualities missing from this list, there will be an opportunity for you to mention this in a written question at the end of the survey.

	1	2	3	4	5
Ease of use					
Maintainability					
Memory use					
Portability					
Correctness / Reliability					
Safety					
Security					
Speed					
Verifiability					

Table 3.3: Rating Question Sample

- Fill-in Question
  Are you satisfied with the current process used for SC software development in your group? If not, what could be done to improve the process?

3.3 Survey Process

The survey process is divided into the following four phases.

- Phase 1: Questionnaire design. 37 questions were designed to find the processes that industry and academia follow when developing SC software.

- Phase 2: Survey pilot testing. After finishing the questionnaire, a pilot test was conducted to evaluate the trial survey questionnaire and the procedures. An invitation email including feedback questions was sent to pilot test participants. The questionnaire was updated according to the feedback from the pilot test.

- Phase 3: Online survey. Invitation emails were sent to participants including the link to the online questionnaire. The survey was intended to take 20-30 minutes to complete.

- Phase 4: Survey Data Analysis. Some basic statistics methods were used to analyze the survey data. A survey report is also provided.

Except phase 4, survey data analysis, which will be introduced in Chapter 4 and phase 1, which is already discussed in the previous section, the other two phases are presented in the following sections.

3.3.1 Pilot Test

Pilot tests are used to provide relevant insight, data, and experience as a basis for decisions to accept, improve, or discard parts of all of the tested survey questionnaires and procedures (Foreman, 1991, page 433). The goal of the pilot test is to evaluate the trial survey questionnaire and the procedures and to make any necessary changes.

In the pilot test, 12 candidates from SC fields were selected as samples. Among these candidates, 6 were faculty members, 3 were graduate students at universities, 2 were experts working in industry and 1 was self employed. All candidates work for at least a portion of their time is SC. A standardized questionnaire containing a series of open and closed questions accompanied by a cover letter and feedback questions were distributed to these candidates via email.

All candidates were provided with a deadline (2 weeks) to return completed questionnaires, either by email or by phone. This deadline was met by majority of candidates and in the case of those candidates who did not meet the deadline, follow-up email request were made. A total of 10 of the 12 responses were eventually received.

After pilot testing, the questionnaire was modified according to the feedback from the pilot testing. For example, additional options were added to question 12 and 16; moreover, some definitions, such as software tools, were added to the questions to make the questions more clear. Appendix A.2 Pilot Test Guidelines provides detailed information of this pilot test, which includes the invitation email and feedback questions. Appendix A.3 Pilot Test Report presents an analysis of feedback and recommendations for updating the questions in the survey.

3.3.2 Online Survey

After the survey questions were updated according to the feedback from pilot test participants, the survey was sent to McMaster Research Ethics Board (MREB), an organization at McMaster University that reviews research involving human participants. In response to the suggestions from MREB, the survey invitation email and a few survey questions were modified. After the survey was approved by MREB, the survey was posted online to the public by an open source software package, eSurveysPro. The survey could be accessed via the following link from December 2007 to February 2008:

http://www.eSurveysPro.com/Survey.aspx?id=b67ce1c1-84c2-4c2b-b66d-70db013d8038

Data Collection

Surveyspro provides functions to save all responses in Excel files. The data collection period was from December 2007 to February 2008.

Participant Recruiting

To involve more people in the survey, the following are ways that we used to recruit participants.

1. Advertisements were posted on some SC relevant newsgroups to absorb some people who might be interested in this survey. The invitation email, presented in Appendix A.1, which was sent to participants, is also posted on the newsgroups as an advertisement to recruit participants.

2. Direct emails were sent to companies and organizations. The contact information, especially email address, of SC software practitioners in industry and academia were found on the Internet. From the website of SC companies and organizations, people's contact information was found. An additional approach to finding participants is that the invitation email invites the respondent to forward the email to other people who have experience in developing SC software and might be interested in this study. In this survey, about 400 invitation emails were sent to SC software practitioners.

3. To find personal interest groups, especially the members of the open source community, we found SC open source software, then we found their corresponding software developers.

4. We also encouraged our friends and colleagues that are involved in developing SC software to participate in the survey.

Chapter 4

Survey Data Analysis

In Chapter 3, the goal of the survey was specified, and, to clarify the goal, 31 research issues (RI) were defined. In this chapter, with the RIs in mind, data collected from the survey is analyzed using Microsoft Excel and SurveysPro. The survey data analysis techniques used in this chapter were inspired by several sources, for example, Chromy and Abeyasekera (2003), E-Cology (2003), ORC Macro International (2000), Gy (1998), Foreman (1991), Desu and Raghavarao (1990), and SPSS (1990).

Sampling theory is very important for survey analysis, as it determines the extent that samples represent the target population, also it affects the confidence with which conclusions can be drawn. Therefore, sampling design is addressed first in this chapter. Section 4.1 discusses the sampling approaches that were used in this survey. Moreover, *margin of error (MOE)*, as a statistic to measure a survey's uncertainty, is also summarized in this section. Section 4.2 introduces the data management strategies to deal with raw data before the data analysis. Then, to help readers better understand our survey results, information about target population and characteristics of the respondents is presented in Section 4.3. After that, Section 4.4 provides the survey results for each RI. For some RIs, the difference between industry and academia, as well as distinctions between different size software and different size groups, are also provided. To clarify the results, bar graphs and line graphs are used to present the results for each RI.

4.1 Sampling Design

The term sampling refers to strategies that enable us to pick a subgroup from a larger group and then use this subgroup as a basis to make inferences about the larger group, or to generalize about the population based on observations of the sample. Sampling is a critical factor in any survey design, determining to what extent the survey results allow reliable inferences to be made within acceptable MOE to the population. A sample design should deal with both the selection of individuals to be included in the sample and the process of estimating population values from the sample values. Selection and estimating are interlinked, as selection rules affect the methods of estimating population values and the precision required for population estimates influence the selection rules. The precision needed depends on the general survey aims, and selection depends on the possibility or feasibility of identifying and approaching the members of the target population. In principle, therefore, survey design and sampling design should go hand in hand (Quinx, 2002).

4.1.1   Sampling Strategies

The assessment of population estimates from sample data requires that the sample is 'representative' of the total population. Careful selection can make a sample more or less representative. This is best achieved by *probabilistic sampling*, whereby each individual of the population has a known non-zero probability of being selected, allowing inferences about the population values by means of statistics computed from the sample data. The basic selection method in probability sampling would be *simple random sampling* in the population. *"Simple random sampling"* is defined as the selection of units from a population in such a manner that each of the different samples consisting of the same number of units has the same known probability of being selected. This also implies that each unit in the population has the same probability of selection" (Foreman, 1991, Page 19).

Letting each unit have the same probability of selection may not always be possible or practical. For instance, simple random sampling surveys are often conducted by starting out with a list (known as the "sampling frame") for all units in the population and then a sample is selected based on a randomizing device that gives each individual a chance of selection. In our survey, it is impossible to create such a list that includes all people involved in developing SC software. Our sampling approach involves searching the Internet, which is not a random sampling, but the process was ad hoc and if repeated would likely yield different results, so we assume that it is approximately random, although, strictly speaking, our approach might cause biases and affect the precision of the population estimates.

The sampling technique used in this project is called *snowball sampling*. *Snowball sampling* is a technique for developing a research sample where existing study subjects recruit future subjects from among their acquaintances. Thus the sample group appears to grow like a rolling snowball. In our project, as mentioned in Chapter 3, advertisements were posted on SC relevant newsgroups to help recruit more participants, which can be regarded as snowball sampling. When using snowball sampling, sample members are not selected from a sampling frame, therefore, snowball samples may cause biases.

To improve the precision of estimates of target quantities, some strategies were

38

tried in this project to improve the quality of the samples. ASA (1998) mentions *strati-fied sampling* designs involve defining groups, or strata, based on characteristics known for everyone in the population, and then taking independent samples within each stratum. Such a design offers flexibility, and, depending on the nature of the strata, they can improve the precision of estimates of target quantities (or equivalently, reduce their MOE). In our survey, we divided the target population into two large groups: industry and academia. After that, we, further, divided these two groups into five subgroups. Concretely, academia was divided into two subgroups: universities and research and development institutes; industry was divided into three subgroups: companies who develop in-house software, software vendors and open source developers. Then, sample lists were created for each subgroup independently, which helps our samples better represent the target population.

To use statistics methods, for example confidence interval and MOE, to analyze the survey data, we assume that the sampling that we did for this survey satisfied *simple random sampling*.

4.1.2 Error Analysis

The *margin of error* is a statistic expressing the amount of random sampling error in a survey's results. The larger the MOE, the less confidence one should have that survey results are close to the "true" figures; that is, the figures for the whole population. The MOE is usually defined as the radius of a *confidence interval* for a particular statistic from a survey. A confidence interval gives an estimated range of values that is likely to include an unknown population parameter, the estimated range being calculated from a given set of sample data. Common choices for the confidence level are 90%, 95%, and 99%. Because the sample size in our project is not large, we calculate MOE based on a 90% confidence level.

The laws of probability make it possible for us to calculate intervals of the form: estimate +/- margin of error. Such intervals are sometimes called 90 percent confidence intervals and would be expected to contain the true value of the target quantity at least 90 percent of the time. The formula for the 90% confidence interval using the t-statistic is $p \pm 1.645\sqrt{p(1-p)/n}$, where p is the proportion, n is the sample size, 1.645 is a t-value obtained from a t-table according to the confidence interval (McDonald, 2008). The derivation of the above formula and t-table can be found in many statistics textbooks, so we do not address these in detail here. An example of calculating confidence intervals comes from the data in the survey where 80 of 165 respondents answer that their group size is 2 to 5 people. In this case, the sample proportion for group size (2 - 5 people) would be 48%. However, according to the above formula, there is an MOE of 6%. Therefore, we can draw a conclusion with 90% confidence that the proportion in the whole population having the same intention on the survey data might be from 42% to 54%.

Three things that might affect the MOE are sample size, the type of sampling done and the size of the population (ASA, 1998).

- sample size

  The size of a sample is a crucial actor affecting the MOE. Table 4.1 presents the relationship between survey sample size and MOE in a 90% level of confidence; a MOE calculator, provided by (Steward, 2008), was used to calculate MOE in the table based on different sample sizes.

Survey Sample Size	Margin of Error Percent
2,000	1.3
1,000	1.9
500	2.6
200	4.2
100	5.9
50	8.3

Table 4.1: Survey Sample Size and Margin of Error Percent

In our survey, the number of respondents is from 110 to 168. In terms of the above table, generally speaking, the MOE for our survey overall should be between 4% and 6% based on 90% confidence interval. This table provides a sense of how our samples represent the whole population. To provide more precise analysis, in the following survey, the MOE will be calculated, where appropriate, for each individual question. Please note that, MOE values are typically calculated for surveys overall but they also should be calculated again when a subgroup of the sample is considered. For example, when we compared the difference in group size between academia and industry, the MOE should be calculated again according to subgroups. In this case, 72 for industry and 78 for academia instead of previous 165 that was used as the overall sample size. Therefore, MOE value (10% and 9%), which are larger than the previous MOE (6%), are obtained. So, the MOE for a subgroup is usually larger than the value for the whole group because the number of respondents in a subgroup is smaller than the overall survey sample size.

- sampling type

  Sampling type also affects the MOE. If *probability sampling*, for example *simple random sampling*, is used in sampling, it will be easy to find MOEs in the survey. Our survey, as mentioned earlier, is not strictly *simple random sampling*, but we still assume we analyze the data as if we had used *simple random sampling*. The real sample errors might be greater than the MOE that we obtained, but the numbers still give a sense of how our samples represent the whole population.

- population size

  ASA (1998) mentions that although it is perhaps surprising to some, the size of the population is a factor that generally has little influence on the margin of error. As an example, a sample size of 100 in a population of 10,000 will have almost the same margin of error as a sample size of 100 in a population of 10 million. Therefore, although the target population of our survey is very large, the relatively small sample size should not affect the MOE.

4.2   Data Preparation

SurveysPro stores all raw data in Excel files; therefore, before proceeding with the data analysis, all raw data was downloaded from the SurveysPro website. The following are the steps that were followed to prepare the data for analysis.

Step 1: Delete all empty records from the Excel files.

Empty records are deleted, but all other responses are kept, even if a respondent did not answer all survey questions. All responses are regarded as valid, because the survey allows people to skip questions that they do not feel comfortable answering.

Step 2: Move all the open-ended answers out of the files.

In the survey, most single choice questions and multiple choice questions provide "Other" option to help respondents fill in additional information. The "Other" option was maintained as a possible value for the variables, but the associated text describing "Other" was moved. The text of the answers was kept in another file for later analysis. In addition, the answers to question 32 to 36, which are fill-in questions, were moved to another file.

Step 3: Identify important variables.

In applied statistics, a variable is a measurable factor, characteristic, or attribute of an individual or a system (Statistics Canada, 2008). Variables are critical in any survey studies. The variables that are relevant to the problem under study must be chosen from the vast array of information available. If important variables are excluded from the data file, the results will be of limited value. In this survey, all information obtained is valuable for our research. For single choice questions, each question can be analyzed using one variable. For instance, Question 2 ("How many people in your current group are involved in developing scientific computing software?") helps us to gain information about the size of SC groups. It is a very important variable and we named it *group size*. For multiple choice questions, for each option of the question, we can regard it as a independent variable. For instance, for question 10, "Which of the following Operating System do you use?," there are seven options for this question, which are Linux/Variants, MacOSX, MS-DOS, IBM OS/2 Warp, Unix/Variants, Windows and Others, therefore, seven variables were defined according to the options. The names of the variables are almost the same as the options of the questions.

When we analyze survey data, we not only need to consider variables by themselves, we sometimes also need to consider the relationships between variables. For example, in this survey, *organization type, group size* and *software size* are three significant variables. In many cases, we need to combine these three variables individually with other variables to get more information.

Coding the data for each variable is also very important. To analyze the survey data easily, the data is presented in numbers instead of texts in data files. This is known as coding the data. For example, instead of using "Yes" or "No" as the values for the *project plan* variable, the codes 1 and 0 were used.

4.3 Target Population and Characteristics of Respondents

To help readers better understand our survey results, information about target population and characteristics of respondents are provided in this section.

The target population of this survey was individuals who have experiences in developing SC software in academia or industry and who speak English. Concretely, the sampling frame, or list of individuals, was derived from the people who are working in the following organizations:

- Companies who developed in-house SC software

- Software vendors who produce custom SC software systems or off-the-shelf SC software

- Research and development institutes

- Universities

- Personal interest group, that is people who are interested in developing SC software but not for commercial purpose, for example the open source community

By March 10, 2008, 307 people visited our online survey, but only 168 people participated in this survey. Among the 168 participants, not including the fill-in questions, 110 respondents completed all questions. Two reasons might cause only half of the visitors to participate in the survey. One reason might be the open source package, Surveyspro, which we selected for our survey. Surveyspro is not very stable, as evidenced by several complaints that were received from participants who informed us that the survey crashed while they were completing it. Once the survey program crashes, the participants are lost, because usually people will not do it again. The other reason might be the number of questions. To get enough information for developing SC software, as mentioned in Chapter 3, 36 questions were design for this survey. According to our pilot test result, usually it takes 20 to 25 minutes to finish the survey. From the survey data, we found that the participant number decreased as the question number increased. So, people might not feel very comfortable to reply to so many questions or spend that much time participating in the survey.

The distribution of respondents is illustrated in Figure 4.1. The largest number of respondents were from universities, with 32% of respondents, then, 23% from companies who develop in-house software, 15% from research and development institutes, 14% from software vendor who produce custom software system or off-the-shelf software, 8% from personal interest group, such as open source community, and 8% from other groups, which are not specified in the survey, such as government and other organizations. Our target population, as mentioned in Chapter 3, is people involved in developing SC software in industry and academia. If we regard research and development institutes and universities as academia, and the rest as industry, then, from the above responses distribution, 47% of respondents came from academia, 45% were from industry and 8% from other groups.

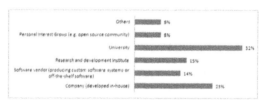

Figure 4.1: Sample Population

4.4 Survey Results

As mentioned in Chapter 1, one of our research goals is to find the current SC software development processes, which we clarified by 31 research issues (RI) in Chapter 3. In this section, four important variables are introduced: *people's education background, people's working experiences, group size* and *software size*. Through the first two variables, the reader can not only obtain the information about SC software practitioners but also better understand the background information about our survey respondents, as discussed in the previous section. The last two variables, *group size* and *software size*, can be combined with many RIs to gain greater insight into the survey data; therefore, they are discussed before presenting the survey results.

The survey results are provided in term of each RI. To simplify the presentation, in the following discussion, only topics for each RI are provided, for detailed information on the RIs please refer to Chapter 3. As mentioned above, *organization type, software size* and *group size* are three important variables in our research; therefore, for some RIs, figures are also provided to present the difference between industry and academia, and software size and group size. The meanings of industry and academia were discussed in Section 4.1.

To clarify the survey results, in most cases, bar graphs are used. The number of respondents for each question is also provided as a part of the title for the corresponding graph. Please note that, for multiple choices questions in the survey, the percentage for each option is presented by number of respondents instead of the number of respondences, hence, the total percentage for all options is usually greater than 100 percent. For example, in Question 10 about operating systems, Table 4.2 presents the number of respondents and the percentage of respondents for each option. From the table, we notice that the total percentage of all options is greater than 100 percent. For single choice questions, the total percentage of all options will be equal to 100 percent.

Options	Number of Respondents that use this OS	Percentage of Respondent
IBM OS/2 Warp	3	2%
Linux / Variants	104	72%
MacOSX	33	23%
MS-DOS	11	8%
Unix / Variants	53	37%
Windows	87	60%
Others	4	3%

Table 4.2: Respondent Distribution for Question 10 (based on 144 respondents)

RI30: Education Background
Figure 4.2 summarizes the responses on the education background of the people who are involved in developing SC software. Overall, people's education backgrounds are very wide, covering many different areas, but, from our survey data, the dominant education background is mathematics, computer science and physics, which are 39%, 30% and 30% respectively.

RI31: Working Experiences
Figure 4.3 and Figure 4.4 demonstrate the working experiences of the respondents. The first figure presents how long each respondent has been working in the SC field and the second one shows how long the respondents have been involved in programming. From these two figures, we see that more than half the respondents have working experiences in SC field of more than 10 years; in addition, more than half the respondents have programming experience of more than 10 years. Moreover, we found that the percentage of people with more than 20 experiences is very high. There may be a potential bias here. Our samples came from an online survey. Maybe people with more experiences are more likely to support our research, so we received more feedback from this type of person.

44

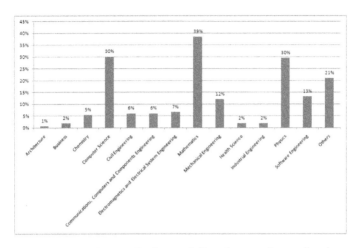

Figure 4.2: Education Background (Based on 166 Respondents)

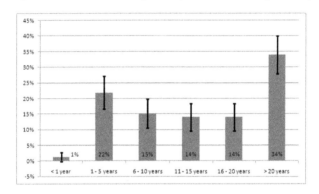

Figure 4.3: Working Experience in SC field (Based on 165 Respondents)

Figure 4.5 summarizes the relationship between software size and people's working experience in the SC field. From the figure, we found that more experienced people seem more likely to be a significant percentage of the developers of large scale software.

Group Size

Figure 4.6 indicates the group size of SC software development groups. We found that SC development groups are usually small groups. From the figure, 42% to 55% of the respondents are in groups of only 2 to 5 people. Also, sometimes (20% to 30% of cases), people might develop SC software by themselves. In this case the software might be for their own usage or for experimental purposes.

Figure 4.7 presents the difference in group size between industry and academia. From the figure, we find that compared with industry, the respondents in academia, have

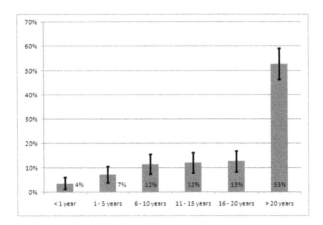

Figure 4.4: Working Experience in Programming (Based on 165 Respondents)

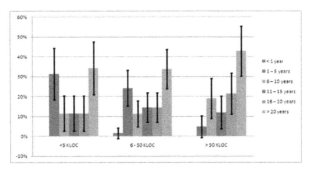

Figure 4.5: The Relationship between People's Working Experience and Software Size (Based on 165 Respondents)

smaller group sizes, with most of their groups being less than 15 people.

Software Sizes

Figure 4.8 demonstrates the software size of SC software. More than 50 percent (56%) of software is less than 20 KLOCS (KLOC = 1000 lines of code). We also notice that large scale SC software (>100 KLOCS) also occupies a high percentage 21%.

4.4.1 Software Development Process

Three questions in the survey, Question 15, 16 and 18, are designed to find information about current SC software development processes. In this section, the associated research issues RI1 to RI3, are discussed.

RI1: Project Plan

In Figure 4.9, 51% of respondents indicate that they will set up a project schedule or a

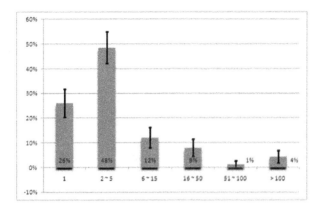

Figure 4.6: Group Size (Based on 165 Respondents)

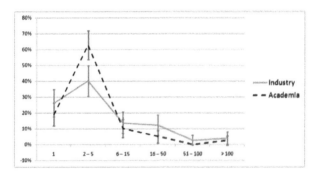

Figure 4.7: Difference in Group Size between Industry and Academia (Based on 165 Respondents)

project plan before developing software, and 49% responses indicate the contrary situation. Error bars in the graph present the MOE, which indicates that the survey data cannot distinguish whether a project plan is more likely than no project plan.

Figure 4.10 shows that larger groups are more likely to set a project schedule or a project plan. Because the sample size of the large groups is very small in our survey, we obtained large MOEs for the largest group size. Even given the large MOEs, the trend seems to follow one's intuition that a larger group will be more likely to have a project plan.

RI2: Process Models

Figure 4.11 illustrates the process models which are currently used in SC development groups. A process that consists of *coding and debugging*, which occupies 58%, is the most frequently used process in SC software development. *Starting from a previous code and modifying it* and *the prototyping model* are also popular approaches, with percentages

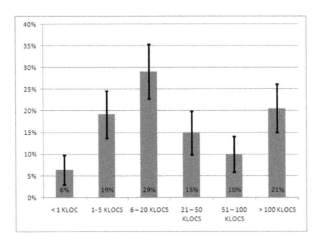

Figure 4.8: Software Sizes (Based on 141 Respondents)

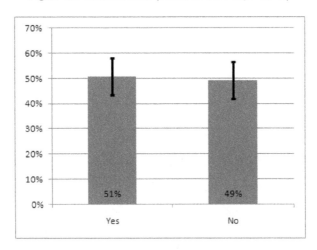

Figure 4.9: Project Plan (Based on 130 Respondents)

of 42% and 41%, respectively. Besides the options which we provided in the survey, respondents also listed some other process models that they are using, such as Extreme Programming, Agile Process, a mix of all the "models" which we provided, or UQDS (ticket and branch-based development) which is described at www.divmod.org.

In industry and academia, we found that there is not much difference in the results. *Coding and debugging* is the most popular approach within both groups, with 45% for both groups. The *coding and debugging* option is also the most popular for all the different sub-population based on group size, with 47% for 1 - 5, 48% for 6 - 50 and

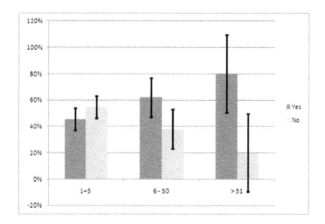

Figure 4.10: Project Plan with Group Size (Based on 130 Respondents)

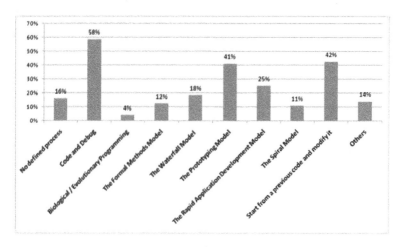

Figure 4.11: Process Models (Based on 132 Respondents)

22% for greater than 50.

Moreover, 24% of respondents only choose the options *No defined process, Code and debug* and *Start from a previous code and modify it*. This means their groups do not adopt a systematic development process and miss the necessary stages, such as requirements specification, during their software development. This may cause significant problems in the software quality, since each stage is essential to software quality. For example, requirements specification guides the entire development process, since, design, coding and testing all depend on it. Missing requirements specification may cause the entire development process to be out of control and thus not meeting the users' needs.

RI3: Time Distribution

Time distribution shows how respondents divide their time during the entire development process. In the survey, we invited participants to provide a time table for a simplified development process, which consists of four stages: requirements specification, design, development and testing. Table 4.3 lists the time distribution for each type of organization. The numbers in the table present the proportion of each stage versus the whole development process. From the table, it is easy to see that the development stage occupies around half the development time in each type of organization. Figure 4.12 presents the time distribution for each type of organization with confidence interval. The trend is also the same when the data is classified by whether the respondent is in academia or industry, as shown in Figure 4.13.

Organization type	Requirements %	Design %	Development %	Testing %
Company	11	16	48	25
Software vendor	13	19	47	20
Research and development institute	14	13	51	22
University	12	16	53	19
Personal Interest Group	13	24	45	18
other	6	23	39	33
All organization	12	17	49	22

Table 4.3: Time Distribution Table (Based on 153 Respondents)

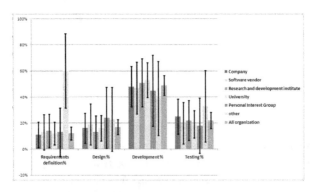

Figure 4.12: Time Distribution with Confidence Interval (Based on 132 Respondents)

Large groups may put more effort into maintainability, since they spend more time on testing instead of coding, which is presented in Figure 4.14. A bias may exist in the data here though because we do not have enough samples from large groups.

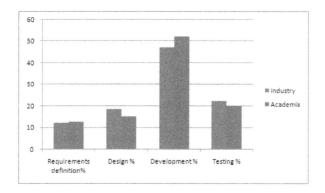

Figure 4.13: Time Distribution in Academia and Industry (Based on 132 Respondents)

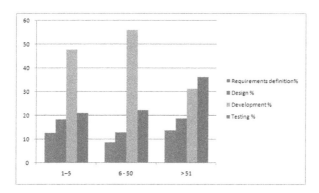

Figure 4.14: Time Distribution between Different Size Groups (Based on 132 Respondents)

4.4.2 Requirements

Requirements specifications are significant to software. Pfleeger and Atlee (2006) mention a survey in their book about the failure of software. The survey analyzes the factors which cause the failure of software projects. Among the factors, the top one is *incomplete requirements* (13%). Other factors such as *unrealistic expectations* (10%), *changing requirements and specifications*(9%) and *lack of planning* (8%) are also related to requirements; therefore, without good requirements specifications, it is hard to complete software, not to mention, to deliver high quality software. Although the above survey is not specific for SC software, it suggests that requirements are important for all kinds of software. Requirements specifications guide the entire development process, that means, designing, coding, testing and maintaining all depend on requirements specifications. Underemphasis on the requirements will reduce the software reliability, reusability, testability and maintainability. Moreover, the final software product needs to meet requirements; therefore, without requirements specification, it is hard to judge the quality of the software. Because of the importance of requirements specifications, five questions were designed in the survey to find whether people use requirements specifications or not and what kind of requirements specifications they are using. In this section, RI4 to RI8, which are relevant to requirements specification, are discussed.

RI4 - RI7: Specifications

Figure 4.15 summarizes the types of specifications currently used in SC development groups. From the figure, only 21% of respondents indicate that there are no specifications in their groups. It seems that some kind of specification is widely used in SC software development. 70% of respondents are using informal specifications. Figure 4.16 further presents the detailed information about specifications. From the figure, 53% and 45% of respondents confirm that they have requirements specifications and design specifications, respectively.

Figure 4.17 shows that semi-formal specification are not often used in SC software development. When comparing the semi-formal specifications approaches, UML is most commonly used. Formal specification is very rarely used, which is illustrated in Figure 4.18. For specification, there is not much difference between industry and academia. Also, distinctions between different size groups and different size software are also not apparent from the data.

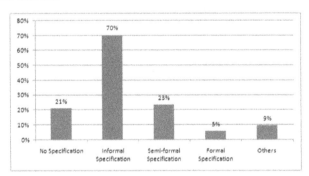

Figure 4.15: Type of Specifications (Based on 128 Respondents)

RI8: Non-Functional Requirements

Figure 4.19 illustrates how important the following software qualities are to SC practitioners: ease of use, maintainability, memory use, portability, correctness, safety, security, speed and verifiability. From the figure, correctness or reliability, ease of use, speed, verifiability and portability are the most important factors to the quality of SC software, especially correctness or reliability. Most respondents, both in industry and academia (shown in Figure 4.20), believe that correctness or reliability is the most significant fac-

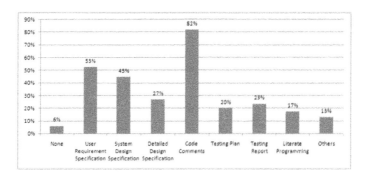

Figure 4.16: Type of Documentation (Based on 116 Respondents)

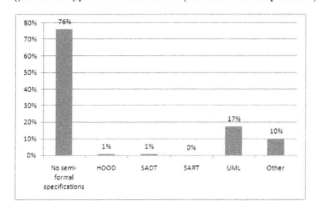

Figure 4.17: Semi-formal Specifications (Based on 121 Respondents)

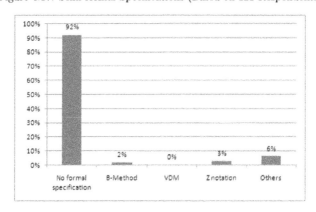

Figure 4.18: Formal Specifications (Based on 114 Respondents)

53

tor that affects the quality of software. This result verifies what we discussed in previous chapters; that is, SC software practitioners care more about correctness than other software qualities. Figure 4.20 shows that industry respondents seem to care more about software's usability, maintainability and safety than the academic respondents.

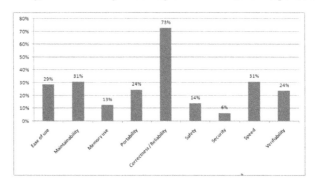

Figure 4.19: Non-functional Requirements (Based on 143 Respondents)

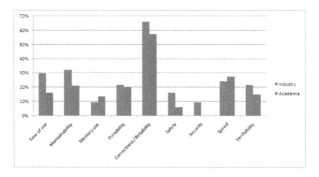

Figure 4.20: Non-function Requirements in Industry and Academia (Based on 143 Respondents)

4.4.3 Design

RI9 to RI13, which are related to software design, are discussed in this section.

RI9: Design Specifications

Figure 4.16 provides information about design specifications. 45% and 27% of respondents indicate that they have system design specification and detailed design specifications, respectively.

RI10: Software Reuse

Figure 4.21 shows that only 2% to 9% of SC software development groups do not reuse software based on 90% confidence interval; therefore, it seems software reuse is already very popular in SC software development. This result is quite different from what we obtained from our literature research, which shows software reuse is poor in SC software development. There might be some bias here that explains the apparent contradiction. One is that a respondent might select one of the reuse options even when the respondent has only used it once or twice. This option may not really be popular in the respondent's group. Another potential source of bias is that respondents might only use a portion of a reuse method but they still select it, which means the respondent did not really use this method during SC software development. For example, a respondent might choose the option sub-system reuse, but, in fact, the respondent only reused the part of the sub-system. However, from the survey results, it is hard to tell these kinds of situations and these situations may cause the percentage that we obtained for software reuse to be higher than it should be. Another reason our survey results differ from Dubios' conclusion that reuse is intended poor in SC Dubios (2002), may be that the respondents of our survey are more like to reuse the software than users in Dubios' lab. All above reasons might cause our survey result to provide a different result than our literature research.

From Figure 4.21, we found that most software reuse exists in the stage of function reuse and module or object reuse. Sub-system reuse and system reuse are not common.

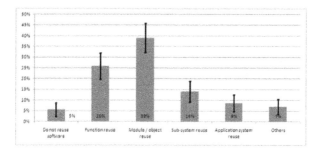

Figure 4.21: Software Reuse (Based on 128 Respondents)

We also tried to find the differences in software reuse between industry and academia, however, in Figure 4.22, after calculating MOEs based on 90% confidence interval, there is no discernible difference between them.

RI11: Mathematical Libraries used in Developing SC Software

In Figure 4.23, 75% of respondents indicate that mathematical libraries are used in developing SC software. BLAS, Netlib (including LAPACK) and GSL are the three most popular libraries. 49% of respondents mention that they use other libraries that are outside the list provided by our survey. Moreover, from the comments from respondents, we found that people also develop in-house mathematical libraries.

RI12: Software Tools

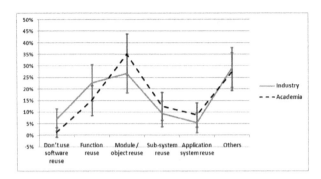

Figure 4.22: Software Reuse in Industry and Academia (Based on 128 Respondents)

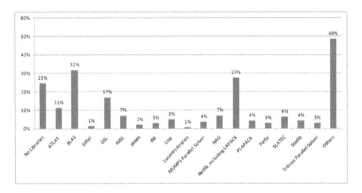

Figure 4.23: Mathematical Libraries used in Developing SC Software (Based on 142 Respondents)

Figure 4.24 summarizes tool usage in SC software development. *Tools for debugging codes* (66%) are very popular in SC develop groups, which is quite reasonable, because from previous survey results, we know that *coding and debugging* is the main approach people are using when developing SC software. *Tools for version control* (69%) is also very popular in SC groups. Moreover, tools for *code generation* and *documentation generation* are also popular with 31% and 37%, respectively. These two percentages are much higher than our expectation, because code generation and documentation generation are very new approaches which we did not expect would be widely used in software development. A likely explanation for the high numbers is that many of the survey participants may have misunderstood our intended meaning of the word generation. Evidence for this exploration can be seen in the responses to Question 34, an open-ended question, where respondents can provide the names of tools that they are using. The answers to Question 34 included Microsoft Word, Latex, and Emacs which suggest that some of the respondents may have confused document and code generation with writing and editing these

56

documents. None of the Question 34 responses included generation tools in the original sense we intended.

Figure 4.25 illustrates the difference in academia and industry for tool use. From this figure, we found that, after calculating margin of error based on 90% confidence interval, 34% to 53% of industry practitioners will use tools for *unit testing*, whereas only 16% to 32% of academia practitioners use unit testing. In other respects, the results do not show any statistically significant difference.

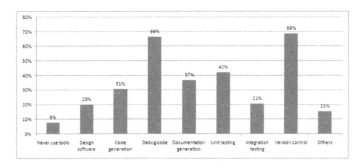

Figure 4.24: Tools (Based on 131 Respondents)

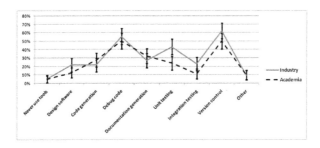

Figure 4.25: Tools in Industry and Academia (Based on 131 Respondents)

4.4.4 Coding

RI14 to RI17 are relative to coding, for example coding standards and tools used in coding.

RI14: Coding Standards

Figure 4.26 presents whether there is coding standards in the respondent's group. For this question, based on 90% confidence interval, 42% to 56% of SC groups have code standard, whereas, 38% to 52% shows the opposite situation. Given the difference between industry

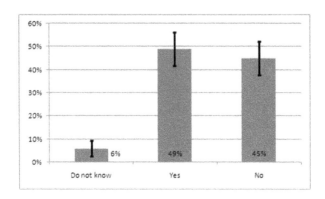

Figure 4.26: Coding Standards (Based on 130 Respondents)

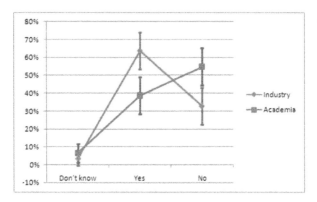

Figure 4.27: Coding Standards in Industry and Academia (Based on 130 Respondents)

and academia, Figure 4.27 shows industry has a greater emphasis on code standard than academia.

RI15 - RI 17: Tools to Generate Codes, Tools to Debug Programs and Tools for Version Control

These three RIs are related to RI12. From Figure 4.24, tools used in code generating, program debugging and version control are 31%, 66% and 69% respectively. According to the comments from respondents, the followings are tools which are commonly used in SC community.

Tools to Generate Codes: Eclipse and Emacs. (Please note that Eclipse and Emacs are editors for editing code not for generating codes.)

Tools to Debug Programs: gdb, ddd, Eclipse and MS Visual Studio built-in debugging tools

Tools for Version Control: SVN and CVS

RI18: Source Code Languages

Figure 4.28 presents the programming languages used for SC software. It is not surprising that C and Fortran are still the most popular programming languages that are used in SC community. We also found that C++, an object oriented programming language, is also often used. Moreover, from the respondents' comments to "Other" option in Question 9, it can be found that some popular programming languages, such as C#, VB.net, which are widely used in developing business applications, are also used by the SC community.

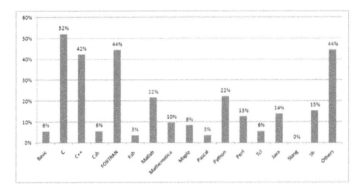

Figure 4.28: Source Code Languages (Based on 144 Respondents)

RI19: Operating Systems

Figure 4.29 summarizes the operating systems that are currently used in developing SC software. From the figure, we see that the dominant operating systems in SC software development are Linux (72%), Windows (60%) and Unix (37%).

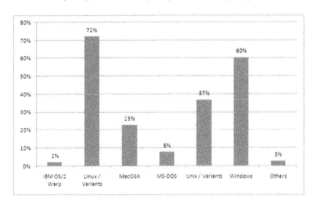

Figure 4.29: Operating Systems (Based on 144 Respondents)

4.4.5 Testing

Testing is very important to develop high-quality software. To obtain information related to testing, such as how to choose test cases, and what kind of validation and verification methods are used in testing, four questions were designed for the survey. These questions address RI20 to RI22.

RI20: Test Cases

Figure 4.30 demonstrates the methods to choose test cases. Requirement specification (68%), boundary value (49%) and logical conditions (43%) are the most important factors when people consider test cases.

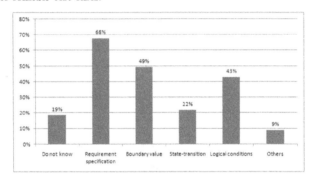

Figure 4.30: Test Cases (Based on 124 Respondents)

RI21: Validation and Verification Methods

Figure 4.31 illustrates the methods which are used in software validation and verification. From the figure, in most cases, the solutions of SC software are verified by: comparing with real world experimental data (75%), comparing with other computational models and simulations (74%) and comparing with closed-form (analytical) solutions (62%).

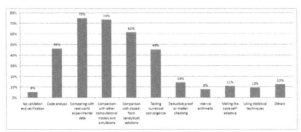

Figure 4.31: Validation and Verification Methods (Based on 125 Respondents)

RI23: Who is in Charge of the Testing Phase

Figure 4.32 shows that usually, in one SC software development group, developers will

60

be in charge of testing, which might cause the problems in the quality of software, since developers are too familiar with the implemenatation's structure and intention. Moreover, developers may have difficulties recognizing the differences between implementation and the required functions or performance. Thus, if, in the development groups, there are professional testers who are independent of the implementation staff and who focus on designing test plans and test cases as well as testing requirements, design components and code, this would be very helpful to improve the quality of the software.

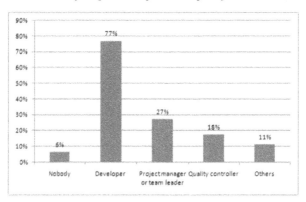

Figure 4.32: Who is in Charge of the Testing Phase (Based on 125 Respondents)

4.4.6 Maintenance

There is only one RI related to maintenance, which is the question related to the lifetime of SC software.

RI25: Software Lifetime

Figure 4.33 indicates the lifetime of SC software. From the figure, we see that the lifetime of this kind of software is very long. Only 4% of the software has a lifetime less than 1 year. More than 70% of SC software will be used for more than 6 year. Moreover, 22% of SC software has a lifetime of more than 20 years. From this, we know that maintainability is very important for SC software.

When comparing the difference between industry and academia about software lifetime, we found from Figure 4.34 that software in academia with a lifetime of 1 to 5 years occupies the highest percentage (31%). We guess the reasons might be that normally in universities, software is developed by students and after the student graduates, the life of software will be over. This theory matches with the fact that graduate students usually are in university less than 5 years. Moreover, software in industry has longer lifetime than academia software. For those software whose lifetime is more than 20 years, industry occupied a much higher percentage than that shown in academia.

61

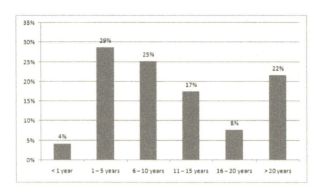

Figure 4.33: Software Lifetime (Based on 143 Respondents)

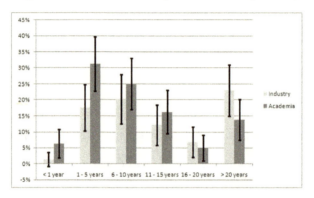

Figure 4.34: Software Lifetime in Industry and Academia (Based on 143 Respondents)

In addition, we found large scale software tends to have a longer life. When taking look at the software whose size is larger than 50 KLOC in Figure 4.35, normally the lifetime of this kind of software is very long.

4.4.7 Documentation

In this section, questions associated with documentation are considered, such as time for documentation updates and the factors that influence good documentation.

RI26: Tools to Generate Documentation
This RI has been discussed in previous RI12 and RI15 to RI 17, therefore, we do not address it again here.

RI27: Speed with which Documentation is Updated
Figure 4.36 shows the time taken to update documentation. From the figure, we are

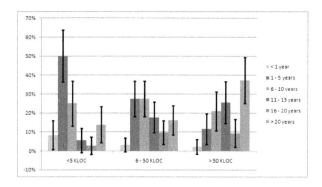

Figure 4.35: Software Lifetime (Based on 143 Respondents)

surprised to find that documentation is changed very often in SC software, especially, based on the 90% confidence interval; 30% to 45% of respondents indicate that the documentation is updated on a daily basis. However, in Figure 4.11 about Process Model, 58% respondents claim that the whole SC software developing process is *code and debug*, which means they do not have documentation during their development process. These two numbers seem inconsistent. We hypothesize that respondents regard code comments as a type of the documentation, thus the frequent updating.

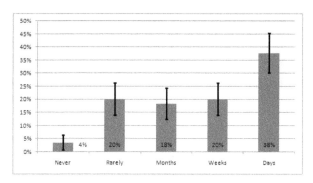

Figure 4.36: Speed with which Documentation is Updated (Based on 114 Respondents)

RI28: Factors of Good Documentation

Figure 4.37 indicates the factors for good documentation. 65% of respondents consider content, i.e. the information that a document contains, as most important. Availability, i.e. ability to retrieve the most current version, and organization, i.e. table of contents, categorized, sub-categorized, etc. are also important.

RI29: Factors Causing Documentation to be out of Sync with the System it Describes

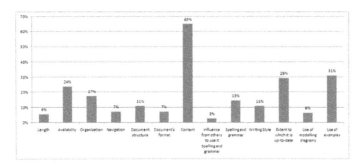

Figure 4.37: Factors of Good Documentation (Based on 109 Respondents)

Figure 4.38 presents the factors that cause the documentation to become outdated. From the figure, time constraints on developers (44%) is the major reason.

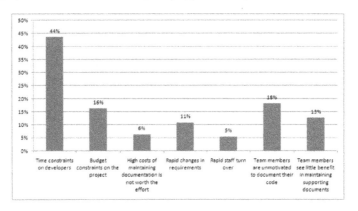

Figure 4.38: Factors Causing Documentation to be out of Sync with the System it Describes (Based on 110 Respondents)

Chapter 5

Proposed Methodologies for Developing ONIS

In Chapter 4, survey results were analyzed to find what SE methodologies are currently used for developing SC software. According to the survey data, potential problems with SC software development were found as follows:

- Systematic development approaches are underemphasized in the SC community, even in large groups. 24% of respondents in our survey mentioned that they do not adopt a systematic development process, for example *coding and debugging* is their entire development process. This code-based approach makes it easy to leave the software undocumented, which may cause severe problems when the system is expanded, or when the original developers are no longer working on the project.

- Although there are requirements specifications in many SC development groups, it seems this stage does not occupy enough of their time. The survey data shows that the average time that a group uses to create requirements specification is about 12% of the total development time.

- Semi-formal and formal specifications are rarely used in SC software development. 72% of respondents indicate that there are no semi-formal specifications in their groups. Furthermore, most respondents (90%) confirmed that they do not use formal specifications. Instead, 70% of respondents admitted that they only use informal specifications, i.e., they use natural language to write their specifications. It is known that nature language is inherently ambiguous, which makes the requirements difficult to validate.

To address the above problems and to show how SE methodologies can be adapted to SC applications, a one-dimensional numerical integration solver (ONIS) is presented as an example. We developed ONIS following two different processes, which could potentially be used by the SC community: i) Parnas' Rational Design Process (PRDP), and ii)

Unified Software Development Process (USDP). The reason why we choose PRDP and USDP are: i) Our version of PRDP is modified specifically for SC problems. Although PRDP is still in the research stage, some efforts, as mentioned before, have been put to make this process fit SC problems. For example, requirements specifications templates for SC problems are already available. Moreover, a couple of examples which use PRDP to develop SC software are also available. This provides some help when SC practitioners would like to use PRDP in their real practice; ii) USDP was chosen because it is a very popular development process and widely used in business applications. The big advantage of USDP is that many resources, such as tools, are available and many of them are open source. We hope to bring this advantage to SC software development. Actually USDP has begun to be used in SC applications as shown by our survey, where 17% of respondents mention that UML, a unified modeling language particularly defined for USDP, is used in their groups. If we use USDP to develop SC software, we can use many existed resources. iii) PRDP and USDP are systematic development process, which can help SC software practitioners to develop SC software using systematic methods. From our survey results, we found that systematic development approach is missing sometimes, PRDP or USDP can help to fix this problem. iv) Both PRDP and USDP use formal methods, which can help SC software practitioners to document precise requirements and design specifications so as to improve the quality of the software.

As mentioned before, ONIS is developed using both PRDP and USDP. In practice, we first used PRDP to design and develop ONIS. After ONIS was finished by PRDP, we redesigned the ONIS using USDP. Using two different processes to develop the same SC software helps us better compare these two processes. In this chapter, the advantages and disadvantages of these two processes are provided. The comparison also can help SC practitioners determine which of the two candidate processes fits their particular situation. Please note that the comparison is based on the experiences during our design and develop ONIS; therefore, there might exist some biases in the conclusion, because the conclusion is based on one case study implemented by one developer. However, the key here is to introduce these two candidate processes and help SC practitioners to more deeply understand these two approaches and possibly convince them to adopt more formal documentation and systematic development approaches to their software. To help SC practitioners better understand these two processes, the complete documents for each of the two processes are provided in Appendices B and C.

There are five sections in this chapter. Section 5.1 briefly introduces ONIS, an SC program that is developed using PRDP and USDP. Section 5.2 summarizes the characteristics of PRDP and USDP and generally compares the advantages and disadvantages of these two processes. In ONIS, requirements specification and design stages of PRDP and USDP are different; therefore, distinctions between PRDP and USDP for these two stages are compared and presented in Section 5.3 and Section 5.4, respectively. According to the advantages and disadvantages of PRDP and USDP, Section 5.5 gives some recommendations to SC practitioners to assist them on making a decision should they want to use one of PRDP or USDP.

5.1 ONIS Introduction

ONIS is a one-dimensional numerical integration solver. The input to ONIS is a function, characteristics of the given function, an interval, and a requested absolute or relative accuracy. ONIS computes the value of numerical integration according to the characteristics of the input function and the given interval. The output is the approximation of the integral, an estimate of the absolute error, the total number of function evaluations that were executed and an error code.

From our survey, we know that mathematical libraries are widely used in SC software development, since 75% of respondents indicate that mathematical libraries are adopted in their software. Using mathematical libraries will benefit the reliability of the developed software, because the popular mathematical libraries have been in use for a long time and their reliability is supported by practice. In ONIS, routines from a FORTRAN library, Quadpack, are used to compute the integration based on the characteristics of an input function. From practice, we found that adopting mathematical libraries in developing SC software not only saves development time, but also improves the reliability of the software.

5.2 USDP vs. PRDP

In Chapter 2, the content and template for USDP and PRDP were discussed. In this section, we present a general comparison of the differences between the two processes with an emphasis on the advantages and disadvantages of each of them.

5.2.1 Common Characteristics of USDP and PRDP

Although USDP and PRDP are different, they do have characteristics in common, as follows:

- High level design
  They both advocate high level design, which means that, in the design stage, they only discuss what the system should do instead of how to do it.

- Platform Independent Design
  Neither USDP or PRDP are specific for one operating system, programming language, or hardware platform. This removes distractions, thus facilitating focusing

on the problems at hand. Also platform independent design can have a long life because it can be applied to future technologies as they arrive. Moreover, platform independent design is not tied to anything implementation-specific; it is generic enough to lend itself to reuse. Please note that USDP can also be used for *platform specific design*, but in ONIS, platform independent design is used.

- Formal methods

  Formal methods are used in both of these processes to avoid ambiguities in specifications; however, informal methods, such as plain English, are also used for specifications. In PRDP, mathematical notation is used to make the process formal; in USDP, the UML constraint language, OCL (Object Constraint Language) based on first-order logic, provides a formal notation for defining complex sets of constraints to define the class invariants, preconditions, postconditions and exceptions.

5.2.2 Differences Between USDP and PRDP

Although there are commonalities between the two processes, there are also many distinctions between them. In particular USDP is fundamentally tied to Object Oriented concepts, such as *class*; therefore, USDP is a natural fit for object-oriented languages and environments such as C++ and Java, but it also can be used to model non-OO applications, for example, FORTRAN applications (OMG, 2008b).

In ONIS, the implementation and testing stages of USDP and PRDP are similar; therefore, in the appendices, only one testing report is provided (Appendix B.4). However, for the requirements specification and the design stage, differences exist, as described in Section 5.3 and 5.4. The following subsections summarize the differences between USDP and PRDP.

5.2.2.1 USDP

In this section, several advantages and disadvantages of USDP are presented, as follow:

- USDP uses a graphical language, which provides advantages over PRDP's text and mathematics based approach. Although PRDP uses a few diagrams, like a graph of the use relation, USDP has a much greater focus on diagrams. UML is the graphical language that is used for USDP. The Object Management Group (OMG) is the body responsible for creating and maintaining the language specifications. They define UML as "a graphical language for visualizing, specifying, constructing, and documenting the artifacts of a software intensive system" (Graham, 2004). UML consists of a large number of different modeling notations, such as *use case diagrams, class diagrams, object diagrams, statecharts, collaboration diagrams, sequence diagrams, activity diagrams,* and *deployment diagrams*. These notations express the different aspects of the system and make the system easier to understand. The big advantage of a graphical language is its intuitive way of expressing ideas; therefore, even those with very little knowledge of programming can participate in the development stages, such as the requirements gathering stage.

However, diagrams also have some disadvantages. For instance, for large systems, too many diagrams might be necessary for expressing all of the different aspects of the system. The large number of diagrams can take a considerable amount of time to build for the entire application. In this case, Graham (2004) suggests that there is no need to tackle the entire UML specification at once. UML has been designed to allow one to use only the sections they need, and to later add the rest incrementally. It is not necessary to use all diagrams defined by UML. In ONIS, only use case diagrams, class diagrams, statecharts, and sequence diagrams were adopted.

- USDP's Model Driven Architecture (MDA) facilitates code and documentation generation. UML forms the foundation of OMG's MDA. A UML model can be either platform-independent or platform-specific. A *Platform-Independent Model* (PIM) represents the system's business functionality and behavior precisely, but does not include technical aspects. From the PIM, MDA-enabled development tools follow OMG-standardized mappings to produce one or more *Platform-Specific Models* (PSM), also in UML. One PSM can be generated for each target platform that the developer chooses. This conversion step can be highly automated. The PSM contains the same information as an implementation, but in the form of a UML model instead of running code. In the next step, the tool generates the running code from the PSM, along with other necessary files, for example, configuration files and makefiles (OMG, 2008b). Moreover, tools are also available for generating documentation automatically. For example, UMLdoc (Gentleware, 2008) is software to generate documentation directly from models. In ONIS, we only create PIM instead of PSM, so resource code and documentation are not automatically generated, but the potential exists for future development in this direction.

- Large resources can be found for USDP. With the popularity of OO concepts, many resources related to USDP are available on the web. For example, a large 710-page PDF file defining UML diagrams is free to download from the OMG website (OMG, 2008a). Moreover, many open source CASE tools, which are used to create diagrams and check diagrams, are also easily found on the web. For ONIS, an open source software, Umlet (Umlet, 2008), was used to create the diagrams.

- OCL helps people without mathematical background write formal specifications. The disadvantage of traditional formal languages is that they are usable to persons with a strong mathematical background, but difficult for the average modeler or developer to use. OCL has been developed to fill this gap. It is a formal language that remains easy to read and write (OMG, 2003, Page 1).

5.2.2.2 PRDP

Compared with USDP, PRDP has some obvious advantages, as follows:

- PRDP fits SC problems better. Our version of PRDP is designed specifically for SC problems; therefore, the templates used in PRDP fits SC problems better than the USDP equivalent. For example, in SC, validation is usually hard, because true answers are difficult to find; hence, we may have to validate the solution via comparison with other programs. In the SRS template, there is a section titled *Solution Validation Strategies*, which can help SC practitioners to validate their SC applications.

- Templates exist to document the stages of the development process. Templates, especially SRS templates, to document general purpose SC software and specific physical problems, are available. Using template not only helps SC practitioners to document the process better, but it also helps people improve the development process, since good templates can remind people what they should do in SC software developments. In some sense, a good template provides a guide for the development team. Templates also exist for USDP. However, most of them are for business applications. No templates exist in USDP specific to SC problems.

- Compared with USDP, PRDP is more abstract. It is easy to find non-abstract sentence like *click the button* in the descriptions for use cases in the examples provided by USDP textbooks. On the other hand, in PRDP, a specification is more abstract because it does not explicitly specify the user interaction. Being abstract, PRDP can better help designers and developers focus on what the system should do and postpone specific design decision later.

- Mathematical notation makes the specification for PRDP more concise than USDP. In USDP, diagrams are used to model different aspects of the system; hence, many diagrams are necessary in the specifications, which make the specifications very long. PRDP uses a mathematical notation. It is known that one of the advantages of mathematics is that it can express ideas more precisely and concisely. Therefore, specifications for PRDP are shorter than those for USDP.

The disadvantage of PRDP is that currently its use is limited; hence, resources relevant to this process are also limited. In fact, this provided one of the reasons for using PRDP for ONIS, since this exercise provides an additional example of the use of PRDP for SC.

5.3 Software Requirements Specification (SRS)

An SRS is significant for the quality of software. Concretely speaking, it can improve the following software qualities for SC software: correctness, usability, maintainability, testability and reusability. For instance, our survey data shows that many SC software has a long lifetime; 70% of SC software will be used for more than 6 year. This makes maintainability of this kind of software very important. Requirements specifications can help maintainers discover and locate errors by comparing the requirements with what the software actually does. In addition, in the survey, when participants were asked which of the factors they consider when they choose test cases, 68% of respondents indicate requirements specification. This shows requirements specification is significant to testing; therefore, unambiguous and validatable software requirements are very helpful for testing the software to improve the testability of the software.

5.3.1 Approach Used in ONIS

To avoid ambiguity and make the SRS as precise as possible, as mentioned previously, a combination of formal methods and informal methods were used in ONIS. Appendix B.1 and C.1 provide the complete SRS for PRDP and USDP, respectively. A good SRS template is very helpful to document the SRS; therefore, a template is chosen for PRDP. The content and history of this template has been discussed in Chapter 2; hence, we do not address it again here. For USDP, no templates existed to document the SRS for SC software; hence, a template which is modified from the SRS of PRDP is provided to document the SRS for USDP.

5.3.2 PRDP vs. USDP for SRS

From appearances, the SRS of PRDP and USDP are very similar, because the template of the SRS for USDP was borrowed from the SRS for PRDP. The difference between these two specifications is that in the SRS for USDP, a *use case model*, illustrated in Figure 5.1, is employed to express the functionality of the system. In addition, USDP adds a *domain model*, presented in Figure 5.2, which is a simple class diagram, which presents the important business concepts to establish the terminologies that are used for writing the descriptions of the use cases. An advantage of using the use case model is that the use case model graphically represents an overview of the functionality of a given system; therefore, it is easy for users to understand the system. The detailed introduction about these two models are provided in Appendix C.1. The disadvantage of this approach is that documenting use case models and domain models need to be supported by tools, like IBM Rational Rose. However, there are now many tools, even open source tools, available to support this process.

71

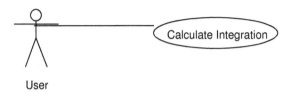

User

Figure 5.1: Use Case Diagram

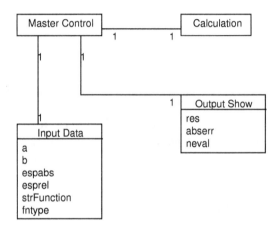

Figure 5.2: Domain Model Diagram

5.4 Software Design

The SRS tells us the problem that the system is intended to solve. In the design stage, the system is shaped and its form (including its architecture) is designed so that it satisfies all requirements - including all nonfunctional requirements and other constraints.

5.4.1 Problems in Software Design

Software reuse is very important, so, in the design stage, we should consider software reusability; that is, the creation and reuse of software building blocks. Such building blocks, often called components or modules, must be cataloged for easy reference, standardized for easy application, and validated for easy integration (Pressman, 2001, Page 121). Therefore, on the one hand, following the principle of *information hiding*, we need to create some modules or components for future usage; on the other hand, we should use existing components (specifications, design, code, and test data) developed for similar past projects. In SC, many mathematical libraries are available. Reusing these libraries not only improves the correctness of the software, but also increases the efficiency of the development process. According to our survey data, we found that 95% of respondents confirm that they reuse software, but with emphasize on function and module reuse. Sub-system and application reuse are rare, with only 14% and 9%, respectively.

Using tools will benefit software design. For example, analysis and design tools help to create models of the system to be built. By performing consistency and validity checking on the models, design tools provide designers or developers with some degree of insight into the analysis representation and help to eliminate errors before they propagate into the design, or worse, into the implementation itself. However, tools use is rare in software design. In the survey, 80% of respondents mention that they do not use tools during design stage. Some respondents said that they only use pen and paper or white boards for design.

Good design specification will help developers better understand the design ideas and decrease mistakes in the implementation. However, according to our survey, more than half of the respondents (55%) indicate that they do not have system design specifications; meanwhile, 73% of respondents state that they do not have detailed design specifications.

5.4.2 Approaches Used in ONIS

In this section, approaches used in ONIS are separately discussed for PRDP and USDP.

5.4.2.1 PRDP

In PRDP, the general idea for software design is: 1) decompose the software into modules, and 2) describe what each module is intended to do and specifying the relationship among the modules. Decomposing the software into modules is actually architecture design, which gives the users the opportunity to view the solution as whole, hiding details that might otherwise distract us; we use a *Module Guide (MG)* to document it. In ONIS, the whole system is decomposed into 5 modules, which are Master Control module, Input Data module, Output Show module, Parser module and Algorithm module. The modules are organized into a hierarchy, as the result of decomposition or abstraction, so that we can investigate the system one level at a time. In the design decomposition, the modules at one level refine those in the level above. As we move to a lower level, we find more detail about each module. Table 5.1 presents the module hierarchy of ONIS. The detail explanation for Table 5.1 is provided in Appendix B.2. Software design also includes the relationship among modules; therefore, the use relation between modules is also provided

Level 1	Level 2
Hardware-Hiding Module	Keyboard Input Module
	Mouse Module
	Screen Display Module
Behavior-Hiding Module	Master Control Module
	Input Data Module
	Output Show Module
	Parser Module
Software Decision Module	Algorithm Module

Table 5.1: Module Hierarchy

in the MG. Figure 5.3 shows the uses hierarchy among modules for ONIS. Parnas (1978) said of two program A and B that A *uses* B if correct execution of B may be necessary for A to complete the task described in its specification. In other words, A *uses* B if there exist situation in which the correct functioning of A depends on the availability of a correct implementation of B.

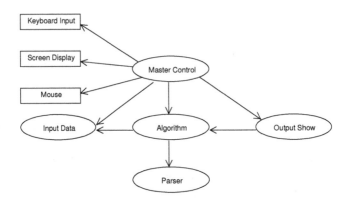

Figure 5.3: Use Hierarchy Between Modules

Detailed software design describes modules and specifying the relationship among the modules. A specification titled *Module Interface Specification (MIS)* is provided to document the design specification at this stage. The MIS clearly defines the inputs and outputs of each module. A complete MIS is provided in appendix B.3. The history and introduction for the templates for MG and the MIS were discussed in Chapter 2.

5.4.2.2 USDP

In the design stage of USDP, an *analysis model* and a *design model* are provided according to the *use case model* in the SRS. The analysis model and the design model can help decompose the whole system into different classes. Figure 5.4 represents the traceability between models.

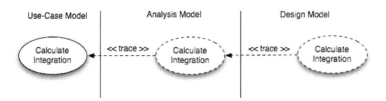

Figure 5.4: Use Case Realizations in the Analysis and Design Models

The analysis model provides a detailed understanding of the requirements. The analysis model provides an overview of the system that may be harder to obtain by studying the results of design or implementation, since too many details are introduced in those stages. The difference between the use case model and the analysis model is provided in Appendix C.2. Figure 5.5 illustrates how the *Calculate Integration* use case is realized by a collaboration with a ≪trace≫ dependency between them, and that four classes participate and play roles in this analysis model. In this analysis model, the Solver Interface is a boundary classes, the Calculation is a control class, and the Algorithm and Parser are entity classes.

In the next step, the analysis model is realized by a design model. Within the design model, use cases are realized by design classes and their objects. In Figure 5.6, four analysis classes participate in realizing the *Calculate Integration* use case: Solver Interface, Calculation, Algorithm and Parser. Also, in the design model, six design classes are refined from analysis classes to adopt to the implementation environment: Input Data, Output Show, Master Control, Algorithm, Parser and Expression are refined from analysis classes. Input Data and Output Show come from the boundary class Solver Interface, which controls the interaction between ONIS and the user, i.e., Input Data helps the user input data to system and Output Show helps the system show the final calculated results. MasterControl comes from the control class, Calculation, if controls the sequence of the system. Algorithm and Parser come from the entity classes with the same names in the analysis model. The functionality of Algorithm is choosing a suitable routine to calculate the integration. Meanwhile, Parser and Expression help to parse the input function and conduct function evaluations.

In USDP, OCL was used to define the class invariants and operations for each class. The OCL expression presented below can be part of a precondition or postcondition that

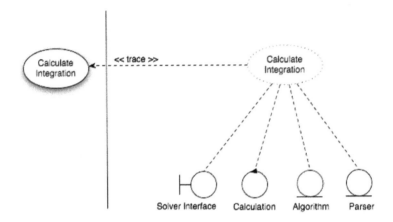

Figure 5.5: Analysis Classes that Participate in a Realization of the Calculate Integration Use Case

is associated with a particular operation. The context declaration in OCL uses the *context* keyword, followed by the type and operation declaration. The constraint is shown by putting the labels 'pre:' and 'post:' before the actual preconditions and postconditions. The following example presents the description for an operation *setUpperbound()* in *InputData* class using OCL. The complete design specification is provided in Appendix C.2.

context InputData :: setUpperbound()
pre: true
post: if $b1 > $ MIN_B and $b1 < $ MAX_B and $b1 >= self.a$
              then $b = b1$
              else ExceptionID = UpperBound_input_invalid
              endif
Assumptions: setUpperbound() is invoked after setLowerbound(), because after the user inputs an upper bound b, the program needs to compare the value of b and a to satisfy the condition $b \geq a$.
Description: setUpperbound() receives a real type upper bound value $b1$ from the keyboard and stores this value in the attribute b.
[Note:] In OCL, the contextual instance *self* is of the type which owns the operation as a feature (IBM, 1997, Page 4). The value of a property on an object that is specified by a dot followed by the name of the property. For instance, *self.a* is the value of the property a on *self*.

A template that combines system design and detail design together for USDP

76

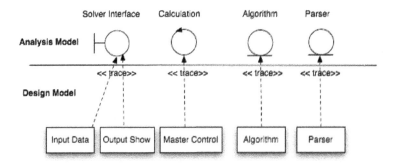

Figure 5.6: Design Classes in the Design Model Tracing to Analysis Classes in the Analysis Model

is implicitly presented through the documentation in Appendix C.2. The template for this software design specification is inspired by the following resources: (Lano, 2005), (Jacobson et al., 1999), (Priestley, 2003) and (OMG, 2003).

5.4.3   USDP vs. PRDP

This section will compare the advantages and disadvantages of USDP and PRDP for the design stage.

5.4.3.1   USDP

The advantages of USDP for design are as follows:

- Dynamic diagrams easily describe the behaviour of the system
  A sequence diagram is used in ONIS to document the main sequence as well as interactions between the objects of ONIS, which is presented by Figure 5.7. Sequence diagrams represent the sequence of the system, and help to understand the application. Moreover, sequence diagrams can easily present the lifetime of object and interactions between objects. In Figure 5.7, the vertical dimension represents time and the messages in an interaction are drawn from top to bottom of the diagram, in the order that they are sent. The dashed line, known as the *lifeline*, indicates the period of time during which objects playing that role actually exist. Messages are shown as arrows leading from the lifeline of the sender of the message to that of the receiver. When the application is not complicated, it is very useful to use a sequence diagram to present the whole system. Compared with PRDP, sequence diagrams really help developers get the whole pictures of the system.

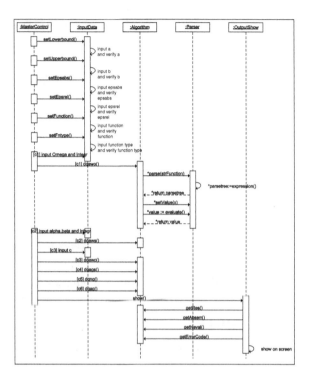

Figure 5.7: Main Sequence of ONIS

- OCL

  OCL is used in the design stage to define the class invariants, preconditions, postconditions and exceptions. Compared with the conventional mathematical notation
  used in PRDP, OCL is also a formal language, but it remains easy to read and write
  especially for people without a mathematical background. Another advantage of
  using OCL is that tools are available to support the OCL language; for example,
  OCL compilers can help validate the constraints that are defined by OCL, which
  improves the correctness of the design, especially when constraints get longer and
  more complicated. It is hard to avoid mistakes in OCL expressions without tools
  support.

- Tools help to improve the maintainability of the specifications and the applications.
  Although building complete models for a given application takes considerable time,
  they are good for future maintenance, especially with the support of tools. If we
  use tools to generate code and design specifications, we can modify the models and
  new code and documentation will be automatically generated.

In the design stage, the disadvantage of USDP is that it takes more time to get familiar with USDP. Compared with PRDP, the process of USDP is relatively complicated. Also, USDP needs more knowledge. For instance, designers and developers need knowledge of modeling, UML and OCL. In addition, given that tools are important in software design in USDP, designers and developers also need to spend time searching and creating relevant tools and becoming familiar with those tools.

5.4.3.2 PRDP

The advantage of PRDP in the design stage is that PRDP is relatively easy to follow. The design process of PRDP is very clear; one is the module decomposition, i.e. decomposing the whole system into modules; the other is module interface specifications, i.e. defining each module in the system. Therefore, Compared with USDP, PRDP is more clearer. Moreover, as mentioned before, in USDP, the number of diagrams are generated in design stage; however, in PRDP, only one diagram is generated, the use hierarchy (Figure 5.3). Obviously, it cannot provide information as detailed as diagrams in USDP; however, it presents the most important relationship between modules, which is the *uses* relationship. The use relationship helps to save time to building unnecessary diagrams, it is also easy to understand. Last but not the least, mathematical notation is an advantage of PRDP, although people with an inadequate background of first order logic might also find it a barrier for use.

5.5 Recommendation

As discussed above, either USDP and PRDP has its advantages and disadvantages. The big advantage of USDP is that USDP is supported by tools. If the users would like to use tools to generate code and documentation automatically, using USDP should be a wise choice, since tools are relative easy to find for USDP. However, comparing with PRDP, USDP is more complicated and it takes more time to learn. Also, it takes more time to build the specifications, since many models and diagrams are needed to express the application using USDP. Moreover, as mentioned in Chapter 2, USDP is a use case driven process, which means the use case is central in this process and all other models have to be built based on the use case model. However, for many SC problems, the use case is very simple. For instance, in ONIS, only one use case is created for this application, which is illustrated in Figure 5.1; therefore, the use case model is not very helpful to express the behaviour of the system. But, without the use case model, USDP cannot proceed, which means the use case model is necessary; therefore, USDP creates some unnecessary diagrams and make the specification loner.

PRDP is designed particular for SC problems, and has templates to follow and its mathematical approach make the specifications more concise; therefore, if the whole develop team has good mathematical background, PRDP might be a better choice.

Chapter 6

Conclusions

Significant quality has been achieved for SC software, but the development of SC software still shows room for improvement. Hence, the goal of our work is to find what qualities of SC software are in most need of improvement, and what kind of SE methodologies can be used to improve the quality of SC software. To achieve our research goal, a survey, which was presented in Chapter 3, including 37 questions, titled Survey on "Developing Scientific Computing Software"', was conducted to obtain the information about the current approaches using in SC software development and how about SC community may respond to new ideas from SE. In terms of the problems in developing SC software, which we obtained through analyzing the survey data, two candidate software development processes were chosen to develop an SC software, ONIS. The goal is to provide an example of how to use SE methodologies to develop an SC software. Through our work, we hope to convince SC practitioners to adopt SE methodologies to develop SC software and to provide useful assistance when they are put into practice.

In this chapter, Section 6.1 provides concluding remarks for this study. Section 6.2 consists of recommendations for future work.

6.1 Concluding Remarks

In this research we designed an SC software survey questionnaire. As mentioned in Chapter 3, a questionnaire with 37 questions were designed specifically for obtaining information about: 1) the current approaches to developing SC software, 2) the software qualities that are in most need of improvement, and 3) the attitudes of the SC community to new ideas from SE. To clarify the goal of the survey, 31 research issues (RI) were defined before designing the survey. After that, to obtain the whole picture of the SC software development process, 37 survey questions were created. These questions covered the whole SC software development process from requirement specification, design, coding, testing and maintenance, using multiple choice multiple answer questions, multiple choice single answer questions, rating questions and fill-in questions. The first three types of questions are suitable for quantitative research. Fill-in questions are intended to get qualitative comments from the respondents.

The following Table 6.1 and Table 6.2 illustrate the relationships between the 31 RIs and the survey questions.

After the survey, we analyzed the survey data. The survey data were collected using an open source software package, Surveyspro, and Microsoft Excel was used to analyze the survey data. An advertisement for the survey was posted on 20 newsgroups to recruit participants; meanwhile, about 400 invitation emails were sent to invite SC practitioners to participate in our survey. Finally, 168 respondences were received. The following summarizes part of our survey findings. The detail survey results were provided in Chapter 4.

- Development Process
  Coding and debugging is the primary process model which is used in SC community, since 58% respondents indicated the entire process of their group is only coding and debugging.

- Requirements Specification (RS)
  RS is adopted in SC software development, however, it does not occupy enough time, i.e. 12% of the total development process. Semi-formal and formal methods are rarely used in current SC community. RS, basically, is documented by natural language. Tools are seldom adopted to generate the requirements document automatically, since, according to the comments from respondents, most of them use Word and Latex to generate RS.

- Design
  SC software practitioners do consider software reuse in the design stage, as shown by more than 90% of respondents confirming that they use some kind of reuse in their software development. As expected, mathematical libraries, as a way of software reuse, are widely used in SC software development.

- Coding
  The time for coding occupies around 50% of the whole development process. C and FORTRAN are still most widely used programming languages in the SC community. Moreover, tools have been adopted in the coding stage, especially for debugging and version control, as shown by the percentages of 66% and 69%, respectively. In terms

Section	NO. of RI	Topic of RI	Associated Questions in Survey
Software Development Process	1	Project Plan	15
	2	Process models	16
	3	Time distribution	23
Requirement	4	Requirement specification	17 and 27
	5	Type of specification	17
	6	Semi-formal specification	18
	7	Formal specification	19
	8	Non-functional Requirement	13
Design	9	Design documentation	27
	10	Software reuse	21
	11	Library used in developing software	12
	12	Tools for software design	22 and 33
	13	Testing plan	25
Coding	14	Coding standard	20
	15	Tools for code generation	22 and 34
	16	Debug tools	22 and 34
	17	Version control tools	22 and 34
	18	Source code and operating system	9
	19	Source code and operating system	10
Testing	20	Test cases	25
	21	Validation and verification methods	24
	22	Testing report	27
	23	Tools for testing	22 and 34
	24	People in charge of testing	26

Table 6.1: Research Issues and Survey Questions 1

Section	NO. of RI	Topic of RI	Associated Questions in Survey
Maintenance	25	Life time of software	14
Documentation	26	Tools to generate documentation	22 and 34
	27	How quickly is documentation updated	28
	28	Factors of good documentation	29
	29	Factors causing documentation to be out of sync with the system it describes	30
People	30	Education background	3
	31	Working experience	5 and 6

Table 6.2: Research Issues and Survey Questions 2

of the respondents comments, automatic code generation is rarely adopted in SC software development.

- Testing

  The special characteristics of SC problems, as mentioned in Chapter 2, the solutions of the SC problems are sometimes unknown; therefore, the solutions of SC software are usually verified by comparing the real world experimental data (75%), comparing with other computation models and simulations (74%) and comparing with closed form (analytical) solutions (62%). Moreover, in testing stage, usually, developers themselves are the people who are responsible for testing.

- Maintenance

  For SC software, usually their lifetime is very long. More than 70% of SC software will be used for more than 6 years, and 20% of SC software has a lifetime of more than 20 years. A long lifetime makes maintenance of this kind of software very important.

In our research, for every RI, differences between industry and academia were compared. In most cases, they are quite similar, but there are still some differences, which are presented as follows:

- Industrial people put more effort on unit testing and are more likely to have a coding standard than academic people

- The group size of academia is smaller than industry. Typically, the group size of academia is less than 15 people.

Our work compared the advantages and disadvantages of PRDP and USDP. To present how to adopt SE methodologies to develop SC software, an example, ONIS was

	PRDP	USDP
Commonalities	High level design Platform independent design Using formal methods	
Differences	Process: SRS → MG → MIS → Implement → Test	Process: Use case model → Analysis model → Design model → Implement → Test
	Do not have tools to support	The whole process needs tools to support
	Formal methods using mathematical notation	Formal methods using OCL language
	Specifications based on textual and mathematical approaches	Specifications based on graphical language UML

Table 6.3: Commonalities and Differences between USDP and PRDP

created with complete documentation using two different development processes, PRDP and USDP. The comparison of the two processes were presented in Chapter 5. Table 6.3 briefly summarizes the key points between the difference between these two processes.

To provide the recommendation for SC software practitioners to help them determine which of the two candidate processes fit their particular situation, the advantages and disadvantages for USDP and PRDP are summarized as follows:

1) USDP

Advantages:

- UML provides a large number of different modeling notations, such as use case diagrams and class diagrams, which can make the system easier to understand.

- Dynamic diagrams such as sequence diagrams can describe the behaviour of the system easily.

- Tools are available to support the whole development process.

- Large resources, such as UML documentation and tools, are available on the web

Disadvantages:

- Too many diagrams inevitably lead to lengthy specifications.

- Considerable knowledge is needed for designers and developers, such as how to use UML, OCL and tools.

2) PRDP

Advantages:

- PRDP was specifically modified for SC problems.

- The whole development process is clear and easy to follow

- Templates, especially templates for requirements specification, are available

- The specifications for PRDP are more concise than that of PRDP, since PRDP use mathematical and textual notation.

Disadvantages:

- The use of PRDP is limited; therefore, the resource of PRDP are also limited.

- People cannot use PRDP without mathematical background especially knowledge of first order logic since mathematical notation is used in PRDP.

6.2   Future Work

The results of our work encourage further research in the field of using SE methodologies to improve the quality of SC software. Some work can be done in the future to promote our research to realize our research goal better, which presents as follows:

- Conduct an improved survey in the future to see the changes in the development of SC software.

- Add more questions to the survey to obtain more information about developing SC software.

- Conduct personal interviews to obtain deeper information that cannot be obtained from a questionnaire.

- Use more reliable survey software to increase the responses of the survey.

- Use more tools in the USDP process, for example to generate documentation and code automatically

- Develop a tools to help maintain consistency between the documents in PRDP.

Bibliography

C. Alexander and S. Ishikawa. A pattern language. In *Oxford: Oxford University Press*, 1977.

Yuri Alexeev, Benjamin A. Allan, and Robert C. Armstrong. Component-based software for high-performance scientific computing. In *Journal of Physics: Comference Series 16, SciDAC 2005*, 2005.

American Statistical Association ASA. What is a margin of error. In *ASA Series: What Is a Survey?*, 1998.

Charles Blilie. Patterns in scientific software: An introduction. In *Computing in Science and Engineering*, 2002.

Ann Christine Catlin. Problem solving environments projects, products, applications and tools. In *http://www.cs.purdue.edu/research/cse/pses/research.html(accessed January,2008)*, 2008.

Org ChangingMinds. Open and closed questions. In *http://changingminds.org/techniques/questioning/open_closed_questions.htm* *(accessed April,2008)*, 2008.

S. Chauvie. Geant4 low energy electromagnetic physics. In *IEEE NSS 2004 Conference, Roma Italy*, 2004.

James R. Chromy and Savitri Abeyasekera. In household surveys in developing and transition countries: Design, implementation and analysis. In *Statistical analysis of survey data*, 2003.

Trevor Cickovski, Thierry Matthey, and Jesus A. Izaguirre. Design patterns for generic object-oriented scientific software. In *Notre Dame Technical Report 2004-29*, 2004.

G. A. P. Cirrone, S. Donadio, S. Guatelli, A. Mantero, B. Mascialino, S. Parlati, A. Pfeiffer, M. G. Pia, A. Ribon, and P. Viarengo. A goodness-of-fit statistical tookit. In *Transactions on Nuclear Science*, 2004.

Alan M. Davis. Software requirements: Analysis and specification. In *Prentice Hall Inc.*, 1990.

Karsten M. Decker and Mark J. Johnson. Application specification and software reuse in parallel scientific computing. In *IEEE Concurrency*, 1998.

M.M. Desu and D. Raghavarao. *Sample size methodology*. Academic Press, Boston, 1990.

E. W. Dijkstra. Chapter notes on structured programming, academic press, london. In *Structured Programming*, 1972.

Paul F. Dubois. Designing scientific components. In *Scientific Programming*, 2002.

Corporation E-Cology. Open source business opportunities for canada's information and communications technology sector: a collaborative fact finding study. In *http://www.e-cology.ca/canfloss/report/CANfloss_Report.pdf(accessed March,2008)*, 2003.

Bo Einarsson. Scientific computing. In *Accuracy and Reliability in Scientific Computing*, 2005.

eSurveysPro. Premium survey services. In *http://www.esurveyspro.com/Default.aspx(accessed May,2008)*, 2008.

E.K. Foreman. *Survey sampling principles.* M. Dekker, New York, 1991.

Andrew Forward. Software documentation - building and maintaining artefacts of communication. In *Master Thesis*, 2002.

M. Fowler. Analysis patterns. In *Analysis Patterns: Reusable Object Models*, 1997.

FSU. Routines for a finite region. In *http://people.scs.fsu.edu(accessed January,2008)*, 2008.

E. Gamma, R. Helm, R. Johnson, and J. Vlissides. Design pattern. In *Design Patterns: Elements of Reusable Object-Oriented Sotware*, 1995.

Gentleware. Poseidon for uml 6.0. In *http://www.gentleware.com/(accessed April,2008)*, 2008.

Elizabeth Graham. Introduction to uml. In *Gentleware Model to Business*, 2004.

GSL. Gsl. In *http://www.gnu.org/software/gsl/(accessed April,2007)*, 2007.

S. Guatelli. Technology transfer from hep computing to the medical field: Overview and application to dosimetry. In *9th Topical Seminar on Innovative Particle and Radiation Detectors Conference*, 2004.

S. Guatelli, B. Mascialino, M. G. Pia, and M. Piergentili. Experience with software process in physics projects. In *American Nuclear Society*, 2005.

Pierre Gy. *Sampling for analytical purposes.* John Wiley, New York, 1998.

Robert R.Korfhage Harley Flanders. Other improper integrals. In *A Second Course in Calculus*, 1974.

Michael T. Heath. Scienfic computing. In *Scientific Computing: An Introductory Survey*, 2003.

Daniel Hoffman and Paul Strooper. Software design, automated testing, and maintenance a practical approach. In *International Thomson Computer Press*, 1999.

IBM. Ocl. In *Object Constraint Language Specification Version 1.1*, 1997.

Ivar Jacobson, Grady Booch, and James Rumbaugh. The unified process. In *The unified Software Development Process*, 1999.

M. Kellner. Introduction to orcan. In *ORCAN Workshop*, 2005.

Diane Kelly and Rebecca Sanders. Assessing the quality of scientific software. In *First International Workshop on Software Engineering for Computational Science and Engineering*, 2008.

J. P. Kenny, S. J. Benson, Y. Alexeev, and J. Sarich. Component-based integration of chemistry and optimization software. In *Computational Chemistry*, 2004.

Jincheol B. Kim, In Soo Ko, and Hyyong Sul. Reengineering and refactoring large-scale scientific programs with the unified process: A case study with osiris pic program. In *Proceedings of EPAC 2004*, 2004.

Konstantin Kreyman and David Lorge Parnas. On documenting the requirements for computer programs based on models of physical phenomena. In *Models 3 August*, 2002.

Lei Lai. Requirements documentation for engineering mechanics software: Guidelines, template and a case study. In *Lei Lai Master Thesis*, 2004.

Kevin Lano. Uml and mda. In *Advanced Systems Design with Java UML and MDA*, 2005.

Kevin Lano, Jose Luiz Fiadeiro, and Luis Andrade. Software design. In *Software Design Using Java2*, 2002.

S. Lefantzi, J. Ray, C. Kennedy, and H. Najm. A component-base toolkit for reacting flow with high order spatial discretizations on structured adaptively refined mesh. In *Comutational Fluid Dynamics: An International Journal*, 2004.

P. Luksch, U. Maier, S. Rathmayer, and M. Weidmann. Sempa: Software engineering methods for parallel scientific applications. In *Software Engineering for Parallel and Distributed Systems*, 1996.

Ruth Malan and Dana Bredemeyer. Functional requirements and use cases. In *Architecture Resources for Enterprise Adventage*, 1999.

A. Mantero. A library for simulated x-ray emission from planetary surfaces. In *IEEE NSS 2004 Conference, Roma Italy*, 2004.

Mathworks. The mathworks accelerating the pace of engineering and science. In *http://www.mathworks.com(accessed February,2007)*, 2007.

John H. McDonald. Confidence limits. In *Handbook of Biological Statistics*, 2008.

Sullivan Mizrahi. Definite integral. In *Calculus and Analytic Geometry*, 1990.

Chris Morris. Some lessons learned reviewing scientific code. In *First International Workshop on Software Engineering for Computational Science and Engineering*, 2008.

NAG. Nag numerical algorithms group. In *http://www.nag.co.uk(accessed February,2007)*, 2007.

Netlib. Slatec. In *http://www.netlib.org/slatec(accessed February,2007)*, 2007.

Netlib. Qag. In *http://www.netlib.org/quadpack/qag.f(accessed January,2008)*, 2008a.

Netlib. Qags. In *http://www.netlib.org/quadpack/qags.f(accessed January,2008)*, 2008b.

Netlib. Qawc. In *http://www.netlib.org/quadpack/qawc.f(accessed January,2008)*, 2008c.

Netlib. Qawo. In *http://www.netlib.org/quadpack/qawo.f(accessed January,2008)*, 2008d.

Netlib. Qaws. In *http://www.netlib.org/quadpack/qaws.f(accessed January,2008)*, 2008e.

Netlib. Qng. In *http://www.netlib.org/quadpack/qng.f(accessed January,2008)*, 2008f.

National Physical Laboratory NPL. Repositories of software. In *http://www.npl.co.uk/server.php?show=ConWebDoc.197(accessed Feburary,2008)*, 2008.

Object Management Group OMG. Ocl specification. In *UML 2.0 OCL Specification*, 2003.

Object Management Group OMG. Uml resource page. In *http://www.uml.org/(accessed April,2008)*, 2008a.

Object Management Group OMG. Intoduction to omg's unified modeling language(uml). In *http://www.omg.org/gettingstarted/what_is_uml.htm(accessed April,2008)*, 2008b.

Inc. ORC Macro International. Evaluation of state-based integrated health information systems. 2000.

Steven Parker. Enabling advanced scientific computing software. In *Software Enabling Technologies for Petascale Science*, 2007.

D. L. Parnas. Designing software for ease of extension and contraction. In *Proceedings of the 3rd Internatinal Conference on Software Engineering*, 1978.

D. L. Parnas and P. C. Clements. A rational design process: How and why to fake it. In *IEEE Transactions on Software Engineering*, 1986.

D. L. Parnas, G.J.K Asmis, and J. Madey. Assessment of safety critical software in nuclear power plants. In *Nuclear Safety*, 1991.

D.L. Parnas, P.C. Clements, and D. M. Weiss. Proceedings of the 7th international conference on software engineering. In *The modular structure of complex system*, 1984.

Shari Lawrence Pfleeger and Joanne M. Atlee. Software engineering. In *Software Engineering Theory and Practice*, 2006.

Roger S. Pressman. Software engineering. In *Software Engineering A Practitioner's Approach (Fifth Edition)*, 2001.

Mark Priestley. Uml. In *Practical Object-Oriented Design With UML*, 2003.

Quadpack. Quadpack numerical integration. In *http://www.csit.fsu.edu/ burkardt/fsrc/quadpack/quadpack.html(accessed Febru-ary,2007)*, 2007.

Research Quinx. Final report. In *Handbook for surveys on drug use among the generation population*, 2002.

Klaus Renzel. Exceptions handling. In *Error Handling for Business Information System*, 2008.

John R. Rice and Ronald F. Boisvert. From scientific software libraries to problem solving environment. In *Computing in Science and Engineering*, 1996.

W. W. Royce. Managing the development of large software system: concepts and techniques. In *IEEE WESTCON, Los Angeles CA: IEEE Computer Society Press*, 1970.

Judith Segal. Models of scientific software development. In *First International Workshop on Software Engineering for Computational Science and Engineering*, 2008.

Emil Sekerinski. Compiler. In *Computer Science 4TB3 Courseware*, 2006.

Mary Shaw and David Garlan. Computer science today: Recent trends and developments. In *Formulations and formalisms in software architecture*, 1995.

W. Spencer Smith. Systematic development of requirements documentation for general purpose scientific computing software. In *Proceedings of the 14th IEEE International Requirements Engineering Conference, RE 2006*, Minneapolis / St. Paul, Minnesota, 2006. URL `http://www.ifi.unizh.ch/req/events/RE06/`.

W. Spencer Smith, Lei Lai, and Ridha Khedri. A new requirements template for scientific computing. In *Proceedings of the First International Workshop on Situational Requirements Engineering Process - Methods, Techniques and Tools to Support Situation - Specific Requirements Engineering Processes*, 2005.

Ian Sommerville. Software engineering. In *Software Engineering Seventh Edition*, 2004.

Inc. SPSS. *SPSS/PC + Statistics 4.0*. SPSS Inc., 1990.

Canada's National Statistical Agency Statistics Canada. Annual survey of software development and computer services. In *The Daily*, 2005.

Canada's National Statistical Agency Statistics Canada. Variable. In *http://www.statcan.ca/english/edu/power/ch8/variable.htm(accessed May,2008)*, 2008.

Associates Steward. Calculate the margin of error based on sample size and other factors. In *http://www.stewardandassociates.net/survey/margin/index.asp(accessed May,2008)*, 2008.

Inc Sun Microsystems. Exceptions. In *http://java.sun.com/docs/books/tutorial/essential/exceptions/dej January,2008)*, 2008.

Umlet. New: Umlet 8, free uml tool for fast uml diagrams. In *http://www.umlet.com/(accessed April,2008)*, 2008.

Inc. Visual Numerics. Visual numerics. In *http://www.vni.com/products/imsl(accessed February,2007)*, 2007.

Manoj Warrier, Shishir Deshpande, and V. S. Ashoka. Scientific computing with free software on gnu / linux howto (accessed january,2008). In *http://tldp.org/HOWTO/Scientific-Computing-with-GNU-Linux/*, 2008.

Wikipedia. Smoothfunction. In *http://en.wikipedia.org/wiki/Smooth_function(accessed April,2007)*, 2007.

Wikipedia. Watts humphrey. In *http://en.wikipedia.org/wiki/Watts_Humphrey(accessed May,2008)*, 2008.

Gregory V. Wilson. Scientists would do well to pick up some tools widely used in the software industry. In *Where's the real bottleneck in scientific computing*, 2006.

Wen Yu. Thesis. In *Improving the Quality of a Parallel Mesh Generation Toolbox by Using Software Engineering Methodologies*, 2007.

Appendix A

Survey on Developing Scientific Computing Software

The followings are questions designed for the survey.

1. Characterization of yourself

Question 1 of 37 (Multiple Choice Single Answer Question)
What is the type of organization where you are currently involved in the development of scientific computing software?

Select one the following:

Company (developed in-house), Software vendor (producing custom software systems or off-the-shelf software), Research and development institute, University, Personal Interest Group (e.g. open source community), Others (identify)

Question 2 of 37 (Multiple Choice Single Answer Question)
How many people in your current group are involved in developing scientific computing software?

Select one the following:

1, 2 - 5, 6 - 15, 16 - 50, 51 - 100, > 100

Question 3 of 37 (Multiple Choice Multiple Answer Question)
What is your education background?

Select all that apply from:

Architecture, Business, Chemistry, Computer Science, Civil Engineering, Communications and Computers and Components Engineering Electromagnetics and Electrical System Engineering, Mathematics, Mechanical Engineering, Health Science, Industrial Engineering, Physics, Software Engineering, Others (identify)

Question 4 of 37 (Multi-Choice Multiple Answer Question))

What is your education background?

Select all that apply from:

Faculty Member, Manager, Project Leader, Researcher, Software Designer, Software Developer, Software Support, Student Technical Writer, Quality Assurance, Other (identify)

Question 5 of 37 (Multi-Choice Single Answer Question)
How long have you been working in the scientific computing field?

Select one the following:

< 1 year, 1 - 5 years, 6 - 10 years, 11 - 15 years, 16 - 20 years, > 20 years

Question 6 of 33 (Multi-Choice Single Answer Question)
How long have you been programming?

Select one the following:

< 1 year, 1 - 5 years, 6 - 10 years, 11 - 15 years, 16 - 20 years, > 20 years

2. Characterization of the scientific computing software that your group is typically involved with developing

Question 7 of 37 (Multi-Choice Multiple Answer Question)
Which of the following fields is your software used in?

Select all that apply from:

Cell Biology, Evolution and Ecology, Molecular and Developmental Genetics, Analytical - Physical Chemistry, Environmental Earth Sciences, General Physics, Inorganic-Organic Chemistry, Solid Earth Sciences, Space and Astronomy, Subatomic Physics, Computing and Information Science, Pure and Applied Mathematics, Statistical Science, Chemical and Metallurgical Engineering, Civil Engineering, Communications, Computers and Components Engineering, Electromagnetics and Electrical System Engineering, Industrial Engineering, Mechanical Engineering, Others (identify)

Question 8 of 37 (Multi-Choice Multiple Answer Question)
What types of scientific computing software are you involved in developing?

Select all that apply from:

Fast Fourier Transform, Interpolation, Linear Solver, Linear Least Squares, Mesh Generation, Numerical Integration, Optimization, Ordinary Differential Equations (ODE) Solver, Random Number Generator, Partial Differential Equations (PDE) Solver, Stochastic Simulation, Solving Eigenvalues, Solving Nonlinear Equations, Others (identify)

Question 9 of 37 (Multi-Choice Multiple Answer Question)

What source code language(s) do you use?

Select all that apply from:

Basic, C, C++, Csh (C Shell Programming), FORTRAN, Ksh (Korn Shell Programming), Matlab, Mathematica, Maple, Pascal, Python, Perl, Tcl, Java, Slang, Sh (Bourne shell Programming), Others (identify)

Question 10 of 37 (Multi-Choice Multiple Answer Question)
Which of the following Operating Systems do you use?

Select all that apply from:

IBM OS/2 Warp, Linux / Variants, MacOSX, MS-DOS, Unix / Variants, Windows, Others (identify)

Question 11 of 37 (Multi-Choice Single Answer Question)
Please approximate the size of typical software you develop in KLOCS (KLOC = 1000 lines of code).

Select one of the following:

< 1 KLOC, 1- 5 KLOCS, 6 - 20 KLOCS, 21 - 50 KLOCS, 51 - 100 KLOCS, > 100 KLOCS

Question 12 of 37 (Multi-Choice Multiple Answer Question)
Which of the following libraries do you use to develop scientific computing software?

Select all that apply from:

No Libraries, ATLAS (Automatically Tuned Linear Algebra Software), BLAS, Deal II, Eiffel Numerical/Scientific Library, GSL (GNU Scientific Library), IMSL, JAMA, Java Numerical/Scientific Libraries (JNL), Lisp Numerical/Scientific Libraries, Lucent Libraries, MUMPS Parallel Solver, NAG, Netlib including LAPACK, PLAPACK (Parallel Linear Algebra), PetSc, SLATEC, Statlib, Trilinos Parallel Solver, Others (identify)

Question 13 of 37 (Rating Question)
In your experience, how important is each of the following software qualities to you. Please rate the relative importance of the qualities, with one (1) for the LEAST important items and five (5) for the MOST important. If you feel that there are software qualities missing from this list, there will be an opportunity for you to mention this in a written question at the end of the survey.

Rate for each of the following:

Ease of use, Maintainability, Memory use, Portability, Correctness / Reliability, Safety, Security, Speed, Verifiability

Question 14 of 37 (Multi-Choice Multiple Answer Question)

What is the lifetime of the typical software that your group develops?

Select all that apply from:

< 1 year, 1 - 5 years, 6 - 10 years, 11 - 15 years, 16 - 20 years, > 20 years

3. Methodology Question 15 of 37 (Multi-Choice Single Answer Question)
In your group, do you set up a project schedule or a project plan before developing software?

Select one of the following:

Yes, No

Question 16 of 37 (Multi-Choice Multiple Answer Question) What kind of process model do you use in developing software?

Select all that apply from:

No defined process, Code and Debug, Biological / Evolutionary Programming (The fittest solution survives), The Formal Methods Model (Specify, develop and verify using rigorous mathematical methods), The Linear Sequential Model / Classic Life Cycle / Waterfall Model (sequentially through requirement, design, coding, testing, and maintenance), The Prototyping Model (start with a "quick design" and a prototype), The Rapid Application Development Model (incremental with a short development cycle), The Spiral Model (identify the sub problem with the highest risk, find a solution, repeat), Start from a previous code and modify it, Others (identify)

Question 17 of 37 (Multi-Choice Multiple Answer Question)
What kind of specifications do you currently use to design and document scientific computing software?

Select all that apply from:

No Specification, Informal Specification (in natural language), Semi-formal Specification, Formal Specification, Others (identify)

Question 18 of 37 (Multi-Choice Multiple Answer Question)
Which of the following semi-formal specification approaches do you use?

Select all that apply from:

No semi-formal specifications, HOOD, SADT, SART, UML, Others (identify)

Question 19 of 37 (Multi-Choice Multiple Answer Question)
Which of the following formal specification approaches do you use? A formal specification uses mathematical methods to document at least portions of the requirements and / or the design.

Select all that apply from:

No formal specification, B-Method, VDM, Z notation, Others (identify)

Question 20 of 37 (Multi-Choice Single Answer Question)
In your current group, is there coding standards that whole group needs to follow?

Select one of the following:

Do not know, Yes, No

Question 21 of 37 (Multi-Choice Single Answer Question) In your current group, when you consider software reuse, what level of software reuse do you reach?

Select one of the following:

Do not use software reuse, Function reuse, Module / object reuse, Sub-system reuse, Application system reuse, Others (identify)

Question 22 of 37 (Multi-Choice Multiple Answer Question)
A programming tool or software tool is a program or application that software developers use to create, debug, or maintain other programs and applications. When you develop software, where do you use tools?

Select all that apply from:

Never use tools, Design software, Code generation, Debug code, Documentation generation, Unit testing, Integration testing, Version control, Others (identify)

Question 23 of 37 (Fill-in Question)
Please estimate the respective ratio of time spent in the following phases, in your software development process: (Please ensure that the numbers sum to 100%)
Requirements definition %
Design (Preliminary and detailed) %
Development (detailed design, coding, debugging) %
Testing (unitary, integration, system, acceptance) %

4. Testing

Question 24 of 37 (Multi-Choice Multiple Answer Question)
What method(s) do you use for software validation and verification?

Select all that apply from:

No validation and verification, Code analysis (using techniques and tools to expose bugs), Comparing with real world experimental data, Comparison with other computational models and simulations, Comparison with closed-form (analytical) solutions, Testing numerical convergence (Verify that the error decreases as the discretization size decreases), Deductive proof or model-checking, Interval arithmetic (Using intervals to track the uncertainty in input quantities to an uncertainty in the result), Making the code self-adaptive (Given a target error tolerance from the user, the software can solve the problem on a sequence of grids until the error estimate is small enough), Benchmark tests, Using statistical techniques (for example Bayesian inference techniques), Others (identify)

Question 25 of 37 (Multi-Choice Multiple Answer Question)
When you choose test cases, which of the factors do you consider?

Select all that apply from:

Do not know, Requirement specification, Boundary value (maximum and minimum number limit), State-transition, Logical conditions, Others (identify)

Question 26 of 37 (Multi-Choice Multiple Answer Question) In your group, who is in charge of the testing phase?

Select all that apply from:

Nobody, Developer, Project manager or team leader, Quality controller, Others (identify)

5. Documentation

Question 27 of 37 (Multi-Choice Multiple Answer Question)
What kind of documentation do you use in developing scientific computing software?

Select all that apply from:

None, User Requirement Specification, System Design Specification, Detailed Design Specification, Code Comments, Testing Plan, Testing Report, Literate Programming (a combination of a programming language, with the main idea of treating a program as a piece of literature.), Others (identify)

Question 28 of 37 (Rating Question)
In your experience, when changes are made to a software system, how long does it take for the supporting documentation to be updated to reflect the changes?

Rate for each of the following:

Never, Rarely, Months, Weeks, Days

Question 29 of 37 (Rating Question)
In your experience, how important is each of the following items in helping to create effective software documentation. Please rate the relative importance of the following items, with one (1) for the LEAST important items and five (5) for the MOST important. If you feel that there are items missing from this list, there will be an opportunity for you to mention this in a written question at the end of the survey.

Rate for each of the following:

Length (not too short, not too long), Availability (ability to retrieve the most current version), Organization (table of contents, categorized, sub-categorized, etc), Navigation (internal / external links, references), Document structure (arrangement of text, tables, figures and diagrams), Document's format (i.e. Microsoft Word, Note Pad, Visio, Html, Pdf), Content (the information that a document contains), Influence from management / project leaders / other developers to use it, Spelling and grammar, Writing Style (choice of words, sentence and paragraph structure), Extent to which it is up-to-date, Use of modeling diagrams (UML, SDL, etc), Use of examples (how to extend or customize a feature)

Question 30 of 37 (Rating Question)
How relevant are the following factors in causing software documentation to be out of sync with the system it describes. Please rate the relative importance of the following items, with one (1) for the LEAST important items and five (5) for the MOST important. If you feel that there are items missing from this list, there will be an opportunity for you to mention this in a written question at the end of the survey.

Rate for each of the following:

Time constraints on developers, Budget constraints on the project, High costs of maintaining documentation is not worth the effort, Rapid changes in requirements, Rapid staff turn over, Team members are unmotivated to document their code, Team members see little benefit in maintaining supporting documents

6. Feedback

Question 31 of 37 (Multi-Choice Single Answer Question) Do you wish to receive a synthesis of this survey?

Select one of the following:

Yes, No

Question 32 of 37 (Multi-Choice Single Answer Question)
We would like to contact some of the survey participants for a follow-up phone interview. Would you potentially be interested in being contacted? If you agree, and are selected, you will receive full details of the follow-up interview in your invitation to participate.

Select one of the following:

Yes, No

Question 33 of 37 (Fill-in Question)
If you said yes to question 31 or question 32, or if you do not mind revealing your identity, please fill in the following contact information.

Name:
Position:
Organization:
Email:
Phone:

Question 34 of 37 (Fill-in Question)
If you use tools, please specify which tools you use
For software design:
For code generation:
For debugging:
For testing:
For version control:
For document generation:

Question 35 of 37 (Fill-in Question)
Please provide any additional comments you may have about the following topics:
What are the important software qualities for scientific computing?
What makes effective software documentation?
What factors causes documentation to be out of sync with the software it describes?

Question 36 of 37 (Fill-in Question)
Are you satisfied with the current process used for scientific computing software development in your group? If not, what could be done to improve the process?

Question 37 of 37 (Fill-in Question)
Please provide any remarks you may have in connection with this questionnaire.

Appendix B

Modified Parnas' Rational Design Process

B.1 SRS for ONIS

This section provides the software requirement specification (SRS) for a one-dimensional numerical integration solver (ONIS) using Modified Parnas' Rational Design Process (PRDP).

B.1.1 Introduction

This section gives an overview of the Software Requirements Specification (SRS) for a One-dimensional Numerical Integration Solver (ONIS). First, the purpose of the document is provided. Second, the scope of an ONIS is identified. The final part of this section summarizes the organization of the document.

B.1.1.1 Purpose of the Document

This SRS provides a "black-box" description of a one-dimensional numerical integration solver. The intended audience of the SRS is the development team and ONIS users whose characteristics are specified in section B.1.2.2.

B.1.1.2 Scope of the Software Product

An ONIS can be used as a single application. It also can be a general purpose tool used by other applications. The ONIS documented here is an alone software. The input of the ONIS is a function, characteristics of the given function, an interval, and absolute accuracy requested or relative accuracy requested. The ONIS computes the value of numerical integration according to the characteristics of the input function and the given interval. The output is the approximation to the integral, an estimate of the absolute error, the total number of function evaluations that were executed and an error code.

B.1.1.3 Organization of the Document

This SRS follows the template given by (Smith, 2006). The rest of the document is organized as follows. Section B.1.2 provides the overall description of the system to make the requirements easier to understand. Section B.1.3 contains all the details of system requirements. Section B.1.4 introduces the non-functional requirements. Section B.1.5 lists the solution validation strategies for this software. Other system issues, traceability matrix, a list of possible changes in the requirements, and values of auxiliary constants are provided in Section B.1.6.

B.1.2 General System Description

This section describes the general information about the system. The system context is presented first. Then the characteristics of the potential users are discussed. At the end of this section, some system constraints are described.

B.1.2.1 System Context

Figure B.1 shows the context for ONIS. A circle represents an external entity outside the system, a user in this case. The rectangle is the system itself. Arrows represent the data flows between them. The "input" is a function, characteristics of the given function,

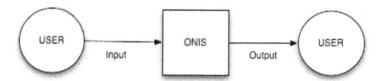

Figure B.1: System Context Diagram

an interval, absolute accuracy requested or relative accuracy requested. The "output" is the approximated value of numerical integration of the given function, an estimate of absolute error, the number of function evaluations and an error code. The function of the ONIS is generating "Output" from "Input."

B.1.2.2 User Characteristics

The target user groups of ONIS are those who are involving in the numerical integration. Anyone who has the following characteristics should be qualified to use this system:

1. Possess an education level that is equivalent to a first or second year university students in science or engineering.

2. Complete university or college first-level calculus course.

3. Complete high school computer related courses.

B.1.2.3 System Constraints

This system will be implemented on Windows or Mac OS environment.

B.1.3 Specific System Description

This section describes the system requirements in detail. After the problem is clearly stated, some solution characteristics are specified. Subsection "Background Overview" provides some background information of the system, and subsection "Terminology Definition" illustrates a list of related concepts. Following this, subsection "Theoretical Models" presents the mathematic model of numerical integration. Then, subsection "Goal Statements" defines the objective. Finally, subsections "Assumptions", "Data Constraint" and "System behavior" form the major parts of this section.

B.1.3.1 Background Overview

The numerical evaluation of integrals is one of the oldest problems in mathematics. The task is to compute the value of the definite integral of a given function. Much effort has been devoted to techniques for the analytic evaluation of the integrals and several libraries have already existed, for example QUADPACK, NAG, IMSL and SLATEC, to

103

calculate the integral.

B.1.3.2 Terminology Definition

The following definitions are sorted by its occurring orders in this document.

- I: true value of $\int_a^b f(x)\,dx$ whose definition is in B.1.3.4.

- Continuous and discontinuous function (Mizrahi, 1990, page 97): let $y = f(x)$ be a function defined on an open interval. The function is said to be continuous at c, if

  1. f(c) is defined and

  2. $\lim_{x \to c} f(x)$ exists and

  3. $\lim_{x \to c} f(x) = f(c)$

  If any one of these three conditions is not satisfied, then the function is said to be discontinuous at c.

- Smooth (Wikipedia, 2007): If the derivative of $f(x)$ is continuous, then $f(x)$ is said to be C^1. If the kth derivative of $f(x)$ is continuous, then $f(x)$ is said to be C^k. By convention, if $f(x)$ is only continuous but does not have a continuous derivative, then $f(x)$ is said to be C^0. And if the kth derivative of $f(x)$ is continuous for all k, then $f(x)$ is said to be C^∞. In other words C^∞ is the intersection $C^\infty = \bigcap_{k=0}^{\infty} C^k$. Differentiable functions are often referred to as smooth. If $f(x)$ is C^k, then $f(x)$ is said to be k_smooth. Most often a function is called smooth (without qualifiers) if $f(x)$ is C^∞ or C^1, depending on the context.

- Singularity (Harley Flanders, 1974, page 35): a definite integral $\int_a^b f(x)\,dx$, a and b finite, is called singular if $f(x)$ "blow up" at one or more points in the interval $[a, b]$. Examples are:
  $$\int_0^3 \frac{1}{x}\,dx, \quad \int_1^5 \frac{1}{x^2 - 4}\,dx, \quad \int_6^{10} \frac{1}{\ln x - 5}\,dx$$
  The first integrand "blows up" at $x = 0$, the second at $x = 2$, the third at $x = 6$. Such bad points are called singularities of the integrand.

- End point singularity: an integral $\int_a^b f(x)\,dx$ where $f(x)$ has singularities which occur either at $x = a$ or at $x = b$. These singularities are called end point singularities.

- CT_QAWS: Integrand can be factored as $f(x)=w(x) \times g(x)$, where $g(x)$ is smooth and $w(x)$ shows a singular behavior at the end points, i.e. $w(x)$ has the following format:

  1. $(x-a)^{\alpha} \times (b-x)^{\beta}$

  2. $(x-a)^{\alpha} \times (b-x)^{\beta} \times \log(x-a)$

  3. $(x-a)^{\alpha} \times (b-x)^{\beta} \times \log(b-x)$

  4. $(x-a)^{\alpha} \times (b-x)^{\beta} \times \log(x-a) \times \log(b-x)$

  where α, β are real, and α, $\beta > -1$.

- Oscillatory function: A function that exhibits oscillation (i.e., slope changes) is said to be oscillating, or sometimes oscillatory.

- CT_QAWC: $f(x)$ can be expressed as $w(x) \times g(x)$, where $g(x)$ is smooth on $[a, b]$ and $w(x) = 1 / (x-c)$ for some constant c.

- CT_QAWO: $f(x)$ can be expressed as $w(x) \times g(x)$, where $g(x)$ is smooth on $[a, b]$ and $w(x)=\cos(\omega x)$ or $\sin(\omega x)$.

- CT_QNG: $f(x)$ is smooth.

- CT_QAGS: $f(x)$ has end point singularities.

- CT_QAG: $f(x)$ has an oscillatory behavior or nonspecific type, and no singularities.

B.1.3.3 Goal Statements

Given function $f(x)$, characteristics C of the given function, an interval $x \in [a, b]$ $(a \le b)$, absolute accuracy requested *epsabs* or relative accuracy requested *epsrel*, return an approximate value y, where $y \approx I = \int_a^b f(x)\,dx$, an estimate of the absolute error ε_a, the number of function evaluations *neval* and an error code *errorcode*.

B.1.3.4 Theoretical Models

Let f be a real valued function defined on the closed interval $[a, b]$. If the function limit exists, then the number I is called the definite integral of f from a to b and it is denoted by $\int_a^b f(x)\,dx$. That is,

Symbol	Type	Meaning	Use
a	$\mathbb{F}$	lower limit of integration	Input
b	$\mathbb{F}$	upper limit of integration	Input
$f(x)$	$\mathbb{R} \rightarrow \mathbb{R}$	input function	Input
C	Ctype	characteristics of the input function	Input
$epsabs$	$\mathbb{F}$	absolute accuracy requested	Input
$epsrel$	$\mathbb{F}$	relative accuracy requested	Input
y	$\mathbb{F}$	approximation to the integral	Output
ε_a	$\mathbb{F}$	estimate of the absolute error	Output
$neval$	Integer	number of function evaluations	Output
$errcode$	Ecodetype	error information	Output

Table B.1: Input and Output Data

$$\int_a^b f(x)\,dx = \lim_{\|P\|\to 0} \sum_{i=1}^{n} f(u_i)\Delta x_i$$

$I = \lim\limits_{\|P\|\to 0} \sum\limits_{i=1}^{n} f(u_i)\Delta x_i$ means that for any given $\epsilon > 0$, there is a positive number δ so that if P is a partition of $[a,b]$ for which $\|P\| < \delta$, then

$$\left| \sum_{i=1}^{n} f(u_i)\Delta x_i - I \right| < \epsilon$$

for any choice of numbers u_i in the subintervals $[x_{i-1}, x_i]$ of P (Mizrahi, 1990, p.349).

B.1.3.5 Data Definition

In this section, some specific data using for solving the problem and data returned by the system will be defined.

- Characteristics of Input Function (Ctype):
  Ctype = {CT_QAWO, CT_QAWS, CT_QAWC, CT_QNG, CT_QAGS, CT_QAG}

- Error Code Type (Ecodetype):
  Ecodetype = {NORMAL, MAX_EVAL_LIMIT, RNDOFF_ERR, LOC_DIFF, NOT_CONVG, DIVG_INGR, INVALID_INPUT}

Variability	Value of Parameter of Variation
V1 (CT_QAWO)	True
V2 (CT_QAWS)	True
V3 (CT_QAWC)	True
V4 (CT_QNG)	True
V5 (CT_QAGS)	True
V6 (CT_QAG)	True
V7 (Entries for a)	$\{\ x : \mathbb{R} \mid x \in \mathbb{F}\colon \text{MIN\_A} \leq a \leq \text{MAX\_A}\}$
V8 (Entries for b)	$\{\ x : \mathbb{R} \mid x \in \mathbb{F}\colon \text{MIN\_B} \leq b \leq \text{MAX\_B}\}$
V9 (Entries for $epsabs$)	$\{\ x : \mathbb{R} \mid x \in \mathbb{F}\colon 0 \leq epsabs \leq \text{MAX\_EPSABS}\}$
V10 (Entries for $epsrel$)	$\{\ x : \mathbb{R} \mid x \in \mathbb{F}\colon 0 \leq epsrel \leq \text{MAX\_EPSREL}\}$
Source of input	Through the user interface
Encoding of input	Text
Format of input $f(x)$	$f(x)$ represented symbolically using expressions in C libraries
Format of input a	the significant digits of input data which should be no more than MAX_IN_DIG
Format of input b	the significant digits of input data which should be no more than MAX_IN_DIG
Format of input $epsabs$	the significant digits of input data which should be no more than MAX_ERR_DIG
Format of input $epsrel$	the significant digits of input data which should be no more than MAX_ERR_DIG

Table B.2: Variabilities for Input Assumptions

Variability	Value OF Parameter of Variation
Check input a and b?	True (check $a \leq b$ and a, b are always numbers)
Check input $epsabs$?	True (check $epsabs \geq 0$ and $epsabs$ are always numbers)
Check input $epsrel$?	True (check $epsrel \geq 0$ and $epsrel$ are always numbers)
Check characteristics C of the input function?	False (assume users input characteristics of $f(x)$ are the same as the actual characteristics of $f(x)$)
Check the input function $f(x)$?	True (check string of input $f(x)$ is not empty)
Exceptions generated?	True

Table B.3: Variabilities for Calculation

B.1.3.6   Assumptions

Input Assumptions (Table B.2)

Calculation (Table B.3)

Output Assumptions (Table B.4)

B.1.4   Data Constraints

B.1.4.1   System Behaviour

Input Variable Behaviour(IV) (Table B.6):

Output Variable Behaviour(OV) (Table B.7):
n = number of function evaluations.

Variability	Value Parameter of Variation
Destination for output y	To screen
Possible value of output y	$\mathbb{F} \cup \{-\infty, \infty, undef\}$
Format of output y	the significant digits of the result which should be no more than MAX_OUT_DIG.
Destination for output ε_a	To screen
Possible value of output ε_a	$\mathbb{F} \cup \{undef\}$
Format of output ε_a	the significant digits of the result.
Destination for output $neval$	To screen
Possible value of output $neval$	Integer
Format of output $neval$	the significant digits of the result.

Table B.4: Variabilities for Output

Variable	Type	System Constraints
a	$\mathbb{F}$	MIN_A $\leq a \leq$ MAX_A
b	$\mathbb{F}$	MIN_B $\leq b \leq$ MAX_B
$epsabs$	$\mathbb{F}$	$0 \leq epsabs \leq$ MAX_EPSABS
$epsrel$	$\mathbb{F}$	$0 \leq epsrel \leq$ MAX_EPSREL
ε_r	$\mathbb{F}$	$0 \leq \varepsilon_r \leq$ MAX_RELERR
ε_a	$\mathbb{F}$	$0 \leq \varepsilon_a \leq$ MAX_ABSERR
$funcount$	Integer	$0 \leq funcount \leq$ MAX_FUN_COUNT

Table B.5: Data Constraints

Input Variable a	Output(ErrorMsg+=)	Output(Abort=)
a <MIN_A||a >MAX_A	INVALID_INPUT	True
Otherwise	NORMAL	False
Input Variable b	Output(ErrorMsg+=)	Output(Abort=)
b <MIN_B||b >MAX_B	INVALID_INPUT	True
Otherwise	NORMAL	False
Input Variable $epsabs$	Output(ErrorMsg+=)	Output(Abort=)
$epsabs$ <0 ||$epsabs$ >MAX_EPSABS	INVALID_INPUT	True
Otherwise	NORMAL	False
Input Variable $epsrel$	Output(ErrorMsg+=)	Output(Abort=)
$epsrel$ <0 ||$epsrel$ >MAX_EPSREL	INVALID_INPUT	True
Otherwise	NORMAL	False
Input Variable $f(x)$	Output(ErrorMsg+=)	Output(Abort=)
String of input $f(x)$ is empty	INVALID_INPUT	True
Otherwise	NORMAL	False
Input Variable C	Output(ErrorMsg+=)	Output(Abort=)
$C \in$ CType	NORMAL	False
Otherwise	INVALID_INPUT	Ture

Table B.6: Input Variable Behaviour

B.1.5 Non-functional Requirements

Requirement Number	N1
Requirement Name	Input Accuracy
Description	The input data of the system are assumed to be accurate, since the input data are given directly by the user of the system. Assuming no human error, there are no input data error.
Binding Time	Specification time
History	Created-Feb, 2007

Behaviour of Algorithm	y	errcode	ε_a	funcount
$(n >$ MAX_FUN_COUNT$)$	undefined	MAX_EVAL_LIMIT	undefined	undefined
round off error occurred in $f(x)$	undefined	RNDOFF_ERR	undefined	undefined
local difficulty in integrand behavior	undefined	LOC_DIFF	undefined	undefined
$f(x)$ is divergent integral	undefined	DIVG_INGR	undefined	undefined
the algorithm does not converge	undefined	NOT_CONVG	undefined	undefined
normal	$\int_a^b f(x)\,dx$	NORMAL	approx $\lvert y - y_{true} \rvert$	n

Table B.7: Output Variable Behaviour

Requirement Number	N2
Requirement Name	Performance
Description	

- speed (response time): for the function $f(x) = \int_{-999999}^{999999} x^2 + 3\,dx$, the response time of ONIS will compare with other systems, for example Matlab.

- precision: the significant digits of the input data should be no more than MAX_IN_DIG and output data should be no more than MAX_OUT_DIG.

Binding Time	Specification time
History	Created-Feb, 2007
Requirement Number	N3
Requirement Name	Tolerance
Description	True solution is not always known. If true solution can be found, computed solution and true solution can be relatively compared.

for the specific test case $f(x) = \int_{-999999}^{999999} x^2 + 3\,dx$,

$|\frac{y_{calculated} - y_{true}}{y_{true}}| \le tol\_ct$

Moreover, computed solution can be compared with third-party systems, for example Matlab.

$|\frac{y_{thirdpart} - y_{calculated}}{y_{calculated}}| \le tol\_thirdpart$

Binding Time	Specification time
History	Created-Feb, 2007
Requirement Number	N4
Requirement Name	Usability
Description	This system should be easy to learn and use. Users have the characteristics indicated before should take no more than REL_TIME to use this application to compute numerical integration for a test case $f(x) = \int_{-999999}^{999999} x^2 + 3\,dx$.
Binding Time	Specification time
History	Created-Feb, 2007

Requirement Number	N5
Requirement Name	Portability
Description	This system should be easily ported to general personal computers with Windows 2000, Windows XP operating systems or later.
Binding Time	Specification time
History	Created-Feb, 2007

B.1.6 Solution Validation Strategies

Solution validation is very important for every system. The following is the solution validation strategies that can be used in ONIS.

B.1.6.1 Relative Comparison between Computed Solution and True Solution

True solutions of the system can not always be obtained, however, sometimes the true solution of test cases can be obtained by some mathematic methods directly, in this case, we can relatively compare the computed solutions and true solution.

B.1.6.2 Compare Solutions with a Third-party System

Computed solutions also can be compared with a third-party system, for example Matlab and Maple. If the solutions are similar, it shows the solution is probably correct.

B.1.6.3 Interval Arithmetic Method

Using interval arithmetic quadrature software for comparing the solutions. If the solution is inside the interval, that means the solution might be correct. If the solution is outside the interval, the solution could be wrong.

B.1.6.4 Using Specific Test Cases

Using some specific test cases, for example giving a very small or very large a and b, for instance, MIN_A for a and MAX_B for b, to check the solutions.

B.1.7 Other System Issues

This section includes some other supporting information that might contribute to the success or failure of the system development. The following factors are considered:

- Open issues are statements of factors that are uncertain and might make significant different to the system.

- Off-the-shell solutions are existing systems or components bought or borrowed. They could potentially satisfy the requirements.

- Waiting rooms provide a blueprint for how the system will be extended.

B.1.7.1 Open Issues

There is no open issues investigated at this stage.

B.1.7.2 Off-the-Shelf Solutions

- Existing system

  - Matlab
    Matlab is both a powerful computational environment and a programming language that handles complex arithmetic. It is a large software package that has many advanced features built-in, and it has become a standard tool for many working in science or engineering disciplines. MATLAB can find both an indefinite integral (i.e., antiderivative) and a definite integral of a symbolic expression. That assumes an elementary antiderivative exists. If not, MATLAB can compute a very accurate numerical approximation to the definite integral (Mathworks, 2007).

- Existing libraries

  - QUADPACK
    QUADPACK is a library of FORTRAN90 routines, using double precision arithmetic, for estimating integrals. The QUADPACK estimate the integral of a function $f(x)$. There are routines for nonadaptive or adaptive integration, finite, semi-infinite or fully infinite integration regions, integrands with singularities, and integrands that include a factor of $\sin(x)$ or $\cos(x)$ (Quadpack, 2007).

  - NAG
    The NAG (Numerical Algorithms Group) library is a comprehensive collection of functions or routines for the solution of numerical and statistical problems (NAG, 2007).

  - IMSL
    The IMSL Numerical Library provides advanced mathematical and statistical functionality for programmers to embed in applications that are written in one of the most widely used programming environments in use today (Visual Numerics, 2007).

  - SLATEC
    The SLATEC Common Mathematical Subroutine Library is an experiment in resource sharing by the computing departments of several Department of Energy Laboratories. The objective is to cooperatively assemble and install at each site a mathematical subroutine library characterized by portability, good numerical technology, good documentation, robustness, and quality assurance. The result is a portable Fortran mathematical subroutine library of

114

	Goal	Input Assumption	Calculation	Output Assumption
IV	X	X	X	
OV	X			X
N1	X	X		X
N2	X			
N3	X	X		X
N4	X			
N5	X			

Table B.8: Traceability Matrix

over 130,000 lines of code (Netlib, 2007).

– GSL

The GNU Scientific Library (GSL) is a numerical library for C and C++ programmers. It is free software under the GNU General Public License. The library provides a wide range of mathematical routines such as random number generators, special functions and least-squares fitting. There are over 1000 functions in total with an extensive test suite (GSL, 2007).

B.1.7.3 Waiting Room

Here, we list the possible changes that can affect the extension of the system.

- W1: The system can recognize features of the given functions automatically.

B.1.8 Traceability Matrix

The traceability matrix defined in the table gives a big picture of the associations among the goal, assumptions, and requirements. The goal is a general problem. After assumptions are applied, the goal is restricted to the problems that can be solved in this system. In the table, when the item on the top header changes, we go through cells of the column of that item. If there is a check mark (X) in a cell, the requirement on the left header should be changed. Tracking these relations is useful for developing and maintaining the software.
IV: Input Variables
OV: Output Variables
N1: Input Accuracy
N2: Performance
N3: Tolerance
N4: Usability
N5: Portability

B.1.9 Values of Auxiliary Constants

- MAX_IN_DIG: a parameter specifying the maximum number of input significant digits, MAX_IN_DIG = 10.

- MAX_OUT_DIG: a parameter specifying the maximum number of output significant digits, MAX_OUT_DIG = 10.

- REL_TIME: time for users to finish the task of calculate the numerical integration for $f(x) = \int_{-999999}^{999999} x^2 + 3\,dx$ using ONIS and Matlab respectively. This constant is used for testing the usability for the system.

- MIN_A: a parameter specifying the minimum number of lower limit of integration a, MIN_A = -999,999.

- MAX_A: a parameter specifying the maximum number of lower limit of integration a, MAX_A = 999,999.

- MIN_B: a parameter specifying the minimum number of upper limit of integration b, MIN_B = -999,999.

- MAX_B: a parameter specifying the maximum number of upper limit of integration b, MAX_B = 999,999.

- MAX_EPSABS: a parameter specifying the maximum number of input absolute accuracy requested *epsabs*, MAX_EPSABS = 1.

- MAX_EPSREL: a parameter specifying the maximum number of input relative accuracy requested *epsrel*, MAX_EPSREL = 1.

- MAX_FUNCOUNT_DIG: a parameter specifying output significant digits of the maximum number of function evaluations *neval*, MAX_FUNCOUNT_DIG = 5.

- MAX_FUN_COUNT: a parameter specifying the maximum number of function evaluations *neval*, MAX_FUN_COUNT = 10,000.

- tol_ct: a parameter specifying the relative error between computed solution and true solution. For the specific test case $f(x) = \int_{-999999}^{999999} x^2 + 3\,dx$, tol_ct should less than 0.1.

116

- tol_thirdpart: a parameter specifying the relative error between computed solution and third-party systems solution. For the specific test case $f(x) = \int_{-999999}^{999999} x^2 + 3 \, dx$, tol_thirdpart $= 0.1$.

- tol_rel: a parameter specifying the relative error between computed solution and ture solution. For the specific test case $f(x) = \int_{-999999}^{999999} x^2 + 3 \, dx$, tol_ct should less than 0.1.

- NORMAL: normal exit.

- RNDOFF_ERR: occurrence of round off error (makes further improvements of the already reached accuracy impossible).

- MAX_EVAL_LIMIT: maximum number of function evaluations has been achieved.

- LOC_DIFF: local difficulty in integrand behavior.

- DIVG_INGR: divergent integral (or slowly convergent integral).

- NOT_CONVG: the algorithm does not converge.

- INVALID_INPUT: invalid input parameters.

B.2 Module Guide for ONIS

This section provides Module Guide (MG) for ONIS.

B.2.1 Introduction

Having completed the Software Requirement Specification (SRS) in the earlier stages, a module structure for a One-dimensional Numerical Integration Solver(ONIS) based on information hiding has been determined. This document specifies the module structure of ONIS will be used in the following ways (Parnas et al., 1984):

- As a guide for new project members - This document can be a guide for a new project member to easily understand the overall structure of the modules and quickly find the relevant modules they are searching for.

- As the support for maintainers - This document will facilitate the process of the maintainers' understanding when making changes in the system. It is important for maintainers to update the relevant sections in the document to reflect the current design after the changes.

- As a verification check for reviewers - Once a module guide has been written, it can be used to check for various errors, for example inconsistency among modules, feasibility of the decomposition and flexibility of the design.

The rest of the document is organized into four sections. Section B.2.2 introduces the connection between requirements and design. Section B.2.3 and Section B.2.4 lists the anticipated changes and unlikely changes of the software requirements. Section B.2.5 summarize the module decomposition. Section B.2.6 lays out the traceability matrix to check the completeness of the design against the requirement. Section B.2.7 illustrates the use hierarchy of all the modules.

B.2.2 Connection Between Requirements and Design

The design of the system should satisfy the SRS. In this stage, the system is decomposed into modules.

B.2.3 Anticipated Changes

This section lists anticipated changes of the system. These changes can make system more mature in the future. The anticipated changes will also result in the changes of the modules.

- AC1: Algorithms for solving numerical integration.

- AC2: Input Device. The input device can be keyboard, file or memory.

- AC3: Methods to get input data and output data.

- AC4: Format of input function, for example, the format of input function can be a tabular data.

- AC5: The sequence of programs being called.

118

- AC6: The data structure of how to store input function.

- AC7: Constraint of lower and upper limit of input function, for example, a and b can be infinity.

- AC8: Algorithms for parsing the input function.

- AC9: The format of input data, that means the significant digits of the input data can be changed.

- AC10: Output Device. The output device can be screen, file or memory.

B.2.4 Unlikely Changes

The following list the potential changes that were not expected to be happened in the system. Sometimes, changing some of these design decisions may lead to big changes of the design.

- UC1: The goal of the system.

- UC2: Dimension of the integral.

- UC3: The representation of the input function. $f(x)$ is always represented symbolically using expressions in C library.

Level 1	Level 2
Hardware-Hiding Module	Keyboard Input Module
	Mouse Module
	Screen Display Module
Behavior-Hiding Module	Master Control Module
	Input Data Module
	Output Show Module
Software Decision Module	Algorithm Module
	Parser Module

Table B.9: Module Hierarchy

The following is the summarization of the leaf modules.
M1: Keyboard Input Module
M2: Mouse Module
M3: Screen Display Module
M4: Master Control Module
M5: Input Data Module
M6: Output Show Module
M7: Parser Module
M8: Algorithm Module

B.2.5   Module Decomposition

This section gives the secret, service of each module. For the leaf modules, anticipated changes are provided. A module hides a change which called the secret of a module. The module service is the functions that the module provided. The anticipated change describes the possible change in the future.

This section is organized as follows: Section B.2.5.1 provides the module guide for the behavior hiding modules. Section B.2.5.2 shows the module guide for the software decision hiding modules. Section B.2.5.3 gives the module guide for the hardware hiding modules.

B.2.5.1   Behavior Hiding Module

This section focuses on the module guide for behavior hiding modules in the ONIS system.

Module name	Behavior Hiding Module
Secrets	The contents of the required behaviors
Services	This module serves as communication layer between the hardware-hiding module and the software decision module.

Master Control Module

Module name	Master Control Module
Secrets	The calling sequence of modules
Services	This module controls the execution sequence of different modules being called through the system.
Anticipated changes	Since this module works as a mediator in the whole system, the sequence of programs being called might be changed. The expected changes correspond to AC5 in the list of anticipated changes.

Input Data Module

Module name	Input Data Module
Secrets	The algorithm on how to input data from screen, and how to store input data into system.
Services	Input data to system from the screen.
Anticipated changes	Services of inputing data to the system might be changed in the future. The expected changes correspond to AC2 and AC9 in the list of anticipated changes.

Output Show Module

Module name	Output Show Module
Secrets	Output format
Services	Output the data on the screen.
Anticipated changes	Output format might be changed and add. The anticipated changes correspond to AC10 in the list of anticipated changes.

B.2.5.2 Software Decision Module

This section focuses on the module guide for software decision modules in the ONIS system.

Module name	Software Decision Module
Secrets	System decision that include data structure and algorithms used in the system.
Services	This module includes data structure and algorithms used in the system.

Parser Module

Module name	Parser Module
Secrets	Algorithm for parsing the input function and function evaluation.
Services	Parse and evaluate the input function.
Anticipated changes	The algorithms for parsing the input might be changed in the future. The anticipated changes correspond to AC8 in the list of anticipated changes.

Algorithm Module

Module name	Algorithm Module
Secrets	Algorithm for calculating the numerical integration
Services	Calculate the numerical integration.
Anticipated changes	The algorithms for calculating numerical integration might be changed in the future. The anticipated changes correspond to AC1 in the list of anticipated changes.

B.2.5.3 Hardware-Hiding Module

This section focuses on the module guide for hardware-hiding modules in the ONIS system.

Module name	Hardware-Hiding Module
Secrets	The data structure and algorithm used to implement the virtual hardware.
Services	Serves as a virtual hardware used by the rest of the system. This module provides the interface between the hardware and the software.

Keyboard Input Module

Module name	Keyboard Input Module
Secrets	The data structure and algorithms for implementing the interface between the keyboard and the system.
Services	Work as a bridge between system software and user and provide all keyboard events that system software needs to respond.
Anticipated changes	Other keyboard event might be added to the system.

Mouse Module

Module name	Mouse Module
Secrets	The data structure and algorithms for implementing the interface between the mouse and the system.
Services	Work as a bridge between system software and user and provide all mouse events that system software needs to respond.
Anticipated changes	Other mouse event might be added to the system based on the need.

Screen display Module

	M1	M2	M3	M4	M5	M6	M7	M8
IV					X			
OV								X
N1					X			
N2				X	X		X	X
N3								X
N4	X	X	X		X	X		
N5	X	X	X		X	X		

Table B.10: Traceability Matrix for Requirement

Module name	Screen display Module
Secrets	Screen Information
Services	Provide screen display functions that system software needs to respond.
Anticipated changes	Other screen display functions might be added to the system based on the need.

B.2.6 Traceability Matrix

Traceability matrix can be used for checking the completeness of current design. This section is organized as follows: Section B.2.6.1 provides the traceability matrix for requirement. Section B.2.6.2 shows the traceability matrix for anticipated changes.

B.2.6.1 Traceability Matrix for Requirement

The traceability matrix in Table B.10 make a connection between modules and requirements. Names and their corresponding numbers of input and output variables of system behavior and non-functional requirement are listed below for convenience.
IV: Input Variable Behavior
OV: Output Variable Behavior
N1: Input Accuracy
N2: Performance
N3: Tolerance
N4: Usability
N5: Portability

B.2.6.2 Traceability Matrix for Anticipated Changes

The traceability matrix in Table B.11 illustrates the relationship between modules and anticipated changes listed in Section B.2.3.

	M1	M2	M3	M4	M5	M6	M7	M8
AC1								X
AC2	X	X			X			
AC3			X		X	X		
AC4					X			
AC5				X				
AC6							X	
AC7					X			
AC8							X	
AC9					X			
AC10						X		

Table B.11: Traceability Matrix for Anticipated Changes

B.2.7 Use Hierarchy between Modules

In this section, use hierarchies between modules are provided. Figure B.2 illustrates the use relation between modules. Squares represent the modules that have developed by OS, and ellipse represent the modules that will implement by the system.

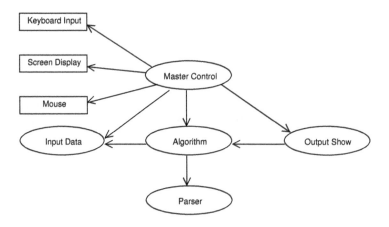

Figure B.2: Use Hierarchy Between Modules

124

B.3 MIS for ONIS

This section provides Module Interface Specification (MIS) for ONIS.

B.3.1 Introduction

This document presents a module interface specification(MIS) for a One-dimensional Numerical Integration Solver (ONIS). It builds based on the Software Requirement Specification and the Module Guide for the ONIS. The rest of the document is organized as follows. Section B.3.2 describes the template used in this MIS document. Section B.3.3 copies the module hierarchy from Module Guide. The rest of sections introduce the Input GUI Module, the Input Data Module, Master Control Module, Output Show Module, Parser Module and Algorithm Module, respectively.

B.3.2 Template Used in MIS

The template used in this MIS is as follows:

- Module Name

- Uses

  - Imported Constants

  - Imported Data Types

  - Imported Access Programs

- Interface Syntax

  - Exported constants

  - Exported Data Types

  - Exported Access Programs

Routine Name	Input	Output	Exception

- Interface Semantics

  - State Variables

  - Invariant

  - Assumptions

  - Access Program Semantics

  - Local Functions

  - Local Data Types

  - Local Constants

  - Considerations

B.3.3 Module Hierarchy

Table B.9 in Appendix B.2 shows the module hierarchy. Master Control Module is the center of ONIS, which controls the sequence of the whole application. Input Data Module provides the user interface to help users input all data, including lower bound, upper bound, function type, function, expected abstract error and expected relative error. After users input all data into ONIS, Input Data Module save all input data. Then, Master Control Module uses Algorithm Module to calculate the value of numerical integration. Algorithm Module picks up different routines according to the characteristics of the input function to calculate numerical integration, meanwhile, it uses Parser Module to parse the input function and do the function evaluation. After Algorithm Module got the final solution, Master Control Module uses Output Show Module to show solutions to the screen.

B.3.4 MIS of Master Control Module

B.3.4.1 Module Name: Master Control Module

B.3.4.2 Uses

Imported Constants
None
Imported Data Types
None
Imported Access Programs
None

B.3.4.3 Interface Syntax

Exported constants
None
Exported Data Types
None
Exported Access Programs

Routine Name	Input	Output	Exception
main			

B.3.4.4 Interface Semantics

State Variables
None
Invariant
None
Assumptions
Master Control Module is the access of this application, so main() is executed before any other routines.
Access Program Semantics
main()

- Description:
  main() controls the sequence of ONIS application. The sequence is as follows:

  1. It uses Input Data Module to get users' input data.
  2. After Input Data Module stored all users' input data, it uses Algorithm module to calculate the numerical integration of the input function.
  3. After Algorithm Module got the final solution, it uses Output Show Module to show results to the screen.

- transition: None

- output: None

Local Constants
None
Considerations
None

B.3.5 MIS of Input Data Module

B.3.5.1 Module Name: Input Data Module

B.3.5.2 Uses

Imported Constants
None
Imported Data Types
None
Imported Access Programs
None

B.3.5.3 Interface Syntax

Exported constants
MIN_A : $\mathbb{F}$
MAX_A : $\mathbb{F}$
MIN_B : $\mathbb{F}$
MAX_B : $\mathbb{F}$
MAX_EPSABS : $\mathbb{F}$
MAX_EPSREL : $\mathbb{F}$
Exported Data Types
Ctype := tuple of {CT_QAWO, CT_QAWS, CT_QAWC, CT_QAGS, CT_QNG, CT_QAG}
Exported Access Programs

Routine Name	Input	Output	Exception
setLowerbound			lowerBound_input_invalid
setUpperbound			upperBound_input_invalid
setEpsabs			absError_input_invalid
setEpsrel			relError_input_invalid
setFunction			function_input_invalid
getLowerbound		$\mathbb{F}$	
getUpperbound		$\mathbb{F}$	
getEpsabs		$\mathbb{F}$	
getEpsrel		$\mathbb{F}$	
getFunction		String	

B.3.5.4 Interface Semantics

State Variables
a: $\mathbb{F}$
b: $\mathbb{F}$
$epsabs$: $\mathbb{F}$
$epsrel$: $\mathbb{F}$
$fntype$: $Ctype$
$strFunction$: $String$

Invariant
None
Assumptions
setUpperbound() is invoked after setLowerbound()

Access Program Semantics
1. setLowerbound()

- Description
  setLowerbound() receives a real type lower bound value $a1$ from the keyboard and stores this value to the attribute a.

- Transition
  $a := a1$

- Exception
  $a < \text{MIN\_A} \vee a > \text{MAX\_A} \Rightarrow \text{lowerBound\_input\_invalid}$

2. setUpperbound()

- Description
  setUpperbound() receives a real type upper bound value $b1$ from the keyboard and stores this value to the attribute b.

- Transition
  $b := b1$

- Exception
  $b < \text{MIN\_B} \vee b > \text{MAX\_B} \Rightarrow \text{upperBound\_input\_invalid}$

3. setEpsabs()

- Description
  setEpsabs() receives a real type absolute accuracy requested value $epsabs1$ from the keyboard and stores this value to the attribute $epsabs$

- Transition
  $epsabs := epsabs1$

- Exception
  $epsabs < 0 \vee epsabs > \text{MAX\_EPSABS} \Rightarrow \text{epsabs\_input\_invalid}$

4. setEpsrel()

- Description
  setEpsrel() receives a real type relative accuracy requested value $epsrel1$ from the keyboard and stores this value to the attribute $epsrel$

- Transition
  $epsrel := epsrel1$

- Exception
  $epsrel < 0 \lor epsrel > \text{MAX\_EPSREL} \Rightarrow \text{epsrel\_input\_invalid}$

5. setFunction($strFunction1 : String$)

- Description
  setFunction() receives a function string $strFunction1$ from the keyboard and stores this string to $strFunction$

- Transition
  $strFunction := strFunction1$

- Exception
  $|\, strFunction\,| = 0 \Rightarrow \text{function\_input\_invalid}$

6. getLowerbound()

- Output
  $\text{out} := a$

7. getLowerbound()

- Output
  $\text{out} := b$

8. getEpsabs()

- Output
  $\text{out} := epsabs$

9. getFunction()

- Output
  out := $strFunction$

Local Functions
None
Local Data Types
None
Local Constants
None
Considerations
None

B.3.6   MIS of Output Show Module

B.3.6.1   Module Name: Output Show Module

B.3.6.2   Uses

Imported Constants
None
Imported Data Types
Uses *AlgorithmModule* Imports *Ecodetype*
Imported Access Programs
None

B.3.6.3   Interface Syntax

Exported constants
None
Exported Data Types
None
Exported Access Programs

Routine Name	Input	Output	Exception
show	$\mathbb{F}$, $\mathbb{F}$, $\mathbb{I}$, *ErrCodeT*	*String, String, String, String, String*	

B.3.6.4   Interface Semantics

Environment Variables
scn: the windows screen
State Variables
None
Invariant
None
Assumptions
None
Access Program Semantics

show(*res*: $\mathbb{F}$, *abserr*: $\mathbb{F}$, *neval*: $\mathbb{I}$, *ier*: *Ecodetype*)

- Transition:

    1. Do the following conversion:
       $str\_approxY = \text{toString}(res)$
       $str\_estAbsError = \text{toString}(abserr)$
       $str\_functionCount = \text{toString}(neval)$
       $str\_errorCode = \text{toStringErr}(ier)$

2. Modify screen *scn* to show *str_approxY*, *str_estAbsError*, *str_functionCount* and *str_errorCode* value on the screen, respectively.

Local Functions

- toString: $\mathbb{F} \longrightarrow String$
  toString(x) return *str_x* where *str_x* $\in String$ and *str_x* is string of x

- toStringErr: *Ecodetype* $\longrightarrow String$
  toStringErr(x) return y where $y \in String$.
  (x=NORMAL $\Longrightarrow y$ = "normal exit" |
  x=MAX_EVAL_LIMIT $\Longrightarrow y$ = "maximum number of function evaluations has been achieved" |
  x=RNDOFF_ERR $\Longrightarrow y$ = "occurrence of roundoff error" |
  x=LOC_DIFF $\Longrightarrow y$ = "local difficulty in integrand behaviour" |
  x=DIVG_INGR $\Longrightarrow y$ = "divergent integral (or slowly convergent integral)" |
  x=INVALID_INPUT $\Longrightarrow y$ = "invalid input parameters")

Local Data Types
None
Local Constants
None
Considerations
None

B.3.7 MIS of Parser Module

B.3.7.1 Module Name: Parser Module

B.3.7.2 Uses

Imported Constants
None
Imported Data Types
None
Imported Access Programs
None

B.3.7.3 Interface Syntax

Exported constants
None
Exported Data Types
Symbol := tuple of {ADD, SUB, MUL, DIV, LBRACK, RBRACK, COMMA, NUM, VAR, SIN, COS, TAN, EXP, LOG, LOG10, POW, SQRT, NUM, EOF, PI, INVALID}
Tflag := tuple of {EXP, VAR, CON}
Expression := tuple of (*typ: Tflag, op: Symbol, opd1: Expression, opd2: Expression, value* : $\mathbb{F}$)
Exported Access Programs

Routine Name	Input	Output	Exception
parse	*String*	*Expression*	
setValue	$\mathbb{F}$		
evaluate	*Expression*	$\mathbb{F}$	

B.3.7.4 Interface Semantics

State Variables
e: set of *Expression*
Invariant
None
Assumptions
parse() is executed first and then setValue(). evaluate() is the last one executed.
Access Program Semantics

1. parse(strFunction: *String*)

- Description:
  The EBNF grammar for input function is as follows:
  expression = ["+" | "-"] term {("+" | "-") term}
  term = factor {("*" | "/") factor}
  factor = number | variable | funct | "(" expression ")"
  funct = COS" | "SIN" | "TAN" | "EXP" | "LOG" | "LOG10" | "POW" | "SQRT"

135

"(" expression ")"

- Transition:
  It calls expression() to parse the input function and generate a parse tree.

2. setValue(val: $\mathbb{F}$)

  - Description: setValue() is used to set the value of each node of parser tree.

  - Transition:
    $e.typ = \text{VAR} \implies e.value = val$ |
    $e.typ = \text{CON} \implies e.value = val$ |
    $e.typ = \text{EXP} \implies$ setValue(val) for $e.op1$ || setValue(val) for $e.op2$ where $e.op2 \neq$ NULL

3. evaluate (function: $Expression$)

  - Transition - output:
    $e.typ = \text{VAR} \implies out := e.value$ |
    $e.typ = \text{CON} \implies out := e.value$ |
    $e.typ = \text{EXP} \implies ($
    $e.op = \text{PLUS} \implies out := \text{evaluate} (e.opd1) + \text{evaluate} (e.opd2)$ ||
    $e.op = \text{MINUS} \implies out := \text{evaluate} (e.opd1) - \text{evaluate} (e.opd2)$ |
    $e.op = \text{TIMES} \implies out := \text{evaluate} (e.opd1) \times \text{evaluate} (e.opd2)$ ||
    $e.op = \text{DIV} \implies out := \text{evaluate} (e.opd1) \div \text{evaluate} (e.opd2)$ ||
    $e.op = \text{SIN} \implies out := \text{sin}(\text{evaluate} (e.opd1))$ ||
    $e.op = \text{COS} \implies out := \text{cos}(\text{evaluate} (e.opd1))$ ||
    $e.op = \text{TAN} \implies out := \text{tan}(\text{evaluate} (e.opd1))$ ||
    $e.op = \text{EXP} \implies out := \text{exp}(\text{evaluate} (e.opd1))$ ||
    $e.op = \text{LOG} \implies out := \text{log}(\text{evaluate} (e.opd1))$ ||
    $e.op = \text{LOG10} \implies out := \text{log10}(\text{evaluate} (e.opd1))$ ||
    $e.op = \text{SQRT} \implies out := \text{sqrt}(\text{evaluate} (e.opd1))$ ||
    $e.op = \text{POW} \implies out := \text{pow}(\text{evaluate} (e.opd1), \text{evaluate} (e.opd2)))$

Local functions

- getSymbol: $String \longrightarrow Symbol$
  getSymbol($s : String$) returns $sym \in Symbol$ where
  $s = \text{"+"} \implies sym = \text{ADD}$
  $s = \text{"-"} \implies sym = \text{SUB}$
  $s = \text{"*"} \implies sym = \text{MUL}$
  $s = \text{"/"} \implies sym = \text{DIV}$
  $s = \text{"sin"} \implies sym = \text{SIN}$
  $s = \text{"cos"} \implies sym = \text{COS}$
  $s = \text{"tan"} \implies sym = \text{TAN}$
  $s = \text{"exp"} \implies sym = \text{EXP}$
  $s = \text{"log"} \implies sym = \text{LOG}$
  $s = \text{"log10"} \implies sym = \text{LOG10}$

s = "pow" $\Longrightarrow$ sym = POW
s = "sqrt" $\Longrightarrow$ sym = SQRT
s $\in$ [1..9] $\Longrightarrow$ sym = NUM
s = "," $\Longrightarrow$ sym = COMM
s = "(" $\Longrightarrow$ sym = LBRACK
s = ")" $\Longrightarrow$ sym = RBRACK
otherwise
sym = INVALID

- expression: $String \longrightarrow$ set of $Expression$
  expression(s: $String$)

  - Tranisition:
    1. Call $getSymbol(s)$ to get $sym \in Symbol$
    2. sym = PLUS $\vee$ sym = MINUS $\Longrightarrow$ call $term(s)$
    3. expression(s) $\equiv$ {$\forall$e: Expression | e.op = PLUS $\vee$ e.op = MINUS $\vee$ e = term(s) : e}

- term: $String \longrightarrow$ set of $Expression$
  term(s: $String$)

  - Tranisition:
    1. Call $getSymbol(s)$ to get $sym \in Symbol$
    2. sym = TIMES $\vee$ sym = DIV $\Longrightarrow$ call $factor(s)$
    3. term(s) $\equiv$ {$\forall$e: Expression | e.op = TIMES $\vee$ e.op = DIV $\vee$ e = factor(s) : e}

- factor: $String \longrightarrow$ set of $Expression$
  factor(s: $String$)

  - Tranisition:
    1. Call $getSymbol(s)$ to get $sym \in Symbol$
    2. $sym \in$ {SIN, COS, TAN, EXP, LOG, LOG10, POW, SQRT} $\Longrightarrow$ call $funct(s)$
    3.factor(s) $\equiv$ {$\forall$e: Expression | e = funct(s) : e}

- funct: $String \longrightarrow$ set of $Expression$
  funct(s: $String$)

  - Tranisition:
    funct(s) $\equiv$ {$\forall$e: Expression | e.op $\in$ {SIN, COS, TAN, EXP, LOG, LOG10, POW, SQRT} $\vee$ expression(s) : e}

Local Data Types
None
Local Constants
None
Considerations
None

B.3.8 MIS of Algorithm Module

B.3.8.1 Module Name: Algorithm Module

B.3.8.2 Uses

Imported Constants
None
Imported Data Types
Uses *Input Data Module* Imports *CType*

Imported Access Programs
Uses *Input Data Module* Imports *getLowerbound*
Uses *Input Data Module* Imports *getUpperbound*
Uses *Input Data Module* Imports *getEpsabs*
Uses *Input Data Module* Imports *getEpsrel*
Uses *Input Data Module* Imports *getFntype*
Uses *Input Data Module* Imports *getFunction*

B.3.8.3 Interface Syntax

Exported constants
Ecodetype = {NORMAL, MAX_EVAL_LIMIT, RNDOFF_ERR, LOC_DIFF, DIVG_INGR, INVALID_INPUT}
Exported Data Types
None
Exported Access Programs

Routine Name	Input	Output	Exception
dqng	*String*, $\mathbb{F}$, $\mathbb{F}$, $\mathbb{F}$,$\mathbb{F}$	$\mathbb{F}$, $\mathbb{F}$, *Ecodetype*, $\mathbb{I}$	calculate_error
dqags	*String*, $\mathbb{F}$, $\mathbb{F}$, $\mathbb{F}$,$\mathbb{F}$	$\mathbb{F}$, $\mathbb{F}$, *Ecodetype*, $\mathbb{I}$	calculate_error
dqags	*String*, $\mathbb{F}$, $\mathbb{F}$, $\mathbb{F}$,$\mathbb{F}$	$\mathbb{F}$, $\mathbb{F}$, *Ecodetype*, $\mathbb{I}$	calculate_error
dqawc	*String*, $\mathbb{F}$, $\mathbb{F}$, $\mathbb{F}$,$\mathbb{F}$	$\mathbb{F}$, $\mathbb{F}$, *Ecodetype*, $\mathbb{I}$	calculate_error
dqawo	*String*, $\mathbb{F}$, $\mathbb{F}$, $\mathbb{F}$,$\mathbb{F}$	$\mathbb{F}$, $\mathbb{F}$, *Ecodetype*, $\mathbb{I}$	calculate_error
dqaws	*String*, $\mathbb{F}$, $\mathbb{F}$, $\mathbb{F}$,$\mathbb{F}$	$\mathbb{F}$, $\mathbb{F}$, *Ecodetype*, $\mathbb{I}$	calculate_error

B.3.8.4 Interface Semantics

State Variables
res: $\mathbb{F}$
abserr: $\mathbb{F}$
ier: *Ecodetype*
neval: $\mathbb{I}$

Invariant
None
Assumptions
None

139

Access Program Semantics
1. dqng(sfun:*String*, a:$\mathbb{F}$, b:$\mathbb{F}$, epsabs:$\mathbb{F}$, epsrel:$\mathbb{F}$)

- Description
  When CType is CT_QNG, dqng will be executed. dqng use *parse*() to parse the function *sfun*; it uses *evaluate*() to execute function evaluation. dqng calculates the integration and gets *res*, *abserr*, *neval* and *ier*.

- Output
  out :=
  $res \in \mathbb{F}$ where *res* is approximate value of $\int_a^b f(x)\, dx$
  $abserr \in \mathbb{F}$ where $abserr = \text{approx}|\, res - y_{true}\,|$
  $neval \in Integer$ where *neval* is approximate function evluation number according to user's expected abstract or relative error.

2. dqags(sfun:*String*, a:$\mathbb{F}$, b:$\mathbb{F}$, epsabs:$\mathbb{F}$, epsrel:$\mathbb{F}$)

- Description
  When CType is CT_QAGS, dqags will be executed. dqags use *parse*() to parse the function *sfun*; it uses *evaluate*() to execute function evaluation. dqags calculates the integration and gets *res*, *abserr*, *neval* and *ier*.

- Output
  out :=
  $res \in \mathbb{F}$ where *res* is approximate value of $\int_a^b f(x)\, dx$
  $abserr \in \mathbb{F}$ where $abserr = \text{approx}|\, res - y_{true}\,|$
  $neval \in Integer$ where *neval* is approximate function evluation number according to user's expected abstract or relative error.

3. dqag(sfun:*String*, a:$\mathbb{F}$, b:$\mathbb{F}$, epsabs:$\mathbb{F}$, epsrel:$\mathbb{F}$)

- Description
  When CType is CT_QAG, dqag will be executed. dqag use *parse*() to parse the function *sfun*; it uses *evaluate*() to execute function evaluation. dqag calculates the integration and gets *res*, *abserr*, *neval* and *ier*.

- Output
  out :=
  $res \in \mathbb{F}$ where *res* is approximate value of $\int_a^b f(x)\, dx$
  $abserr \in \mathbb{F}$ where $abserr = \text{approx}|\, res - y_{true}\,|$
  $neval \in Integer$ where *neval* is approximate function evluation number according to user's expected abstract or relative error.

4. dqawc(sfun:*String*, a:$\mathbb{F}$, b:$\mathbb{F}$, c:$\mathbb{F}$, epsabs:$\mathbb{F}$, epsrel:$\mathbb{F}$)

- Description
  When CType is CT_QAWC, dqawc will be executed. dqawc use *parse*() to parse the function *sfun*; it uses *evaluate*() to execute function evaluation. dqawc calculates the integration and gets *res*, *abserr*, *neval* and *ier*.

- Output
  out :=
  res $\in \mathbb{F}$ where *res* is approximate value of $\int_a^b f(x)\,dx$
  abserr $\in \mathbb{F}$ where *abserr* = approx$| \; res - y_{true} \; |$
  neval $\in$ *Integer* where *neval* is approximate function evluation number according to user's expected abstract or relative error.

5. dqawo(sfun:*String*, a:$\mathbb{F}$, b:$\mathbb{F}$, omega:$\mathbb{F}$, integr:$\mathbb{I}$, epsabs:$\mathbb{F}$, epsrel:$\mathbb{F}$)

- Description
  When CType is CT_QAWO, dqawo will be executed. dqawo use *parse*() to parse the function *sfun*; it uses *evaluate*() to execute function evaluation. dqawo calculates the integration and gets *res*, *abserr*, *neval* and *ier*.

- Output
  out :=
  res $\in \mathbb{F}$ where *res* is approximate value of $\int_a^b f(x)\,dx$
  abserr $\in \mathbb{F}$ where *abserr* = approx$| \; res - y_{true} \; |$
  neval $\in$ *Integer* where *neval* is approximate function evluation number according to user's expected abstract or relative error.

6. dqaws(sfun:*String*, a:$\mathbb{F}$, b:$\mathbb{F}$, alfa:$\mathbb{F}$, beta:$\mathbb{F}$, integr:$\mathbb{I}$, epsabs:$\mathbb{F}$, epsrel:$\mathbb{F}$)

- Description
  When CType is CT_QAWS, dqaws will be executed. dqaws use *parse*() to parse the function *sfun*; it uses *evaluate*() to execute function evaluation. dqaws calculates the integration and gets *res*, *abserr*, *neval* and *ier*.

- Output
  out :=
  res $\in \mathbb{F}$ where *res* is approximate value of $\int_a^b f(x)\,dx$
  abserr $\in \mathbb{F}$ where *abserr* = approx$| \; res - y_{true} \; |$
  neval $\in$ *Integer* where *neval* is approximate function evluation number according to user's expected abstract or relative error.

Local Functions
None
Local Data Types
None
Local Constants
None
Considerations
None

B.4 Testing Report for ONIS

This section provides the testing report for ONIS.

B.4.1 Introduction

This document gives an overview of the testing summary for a One-dimensional Numerical Integration Solver(ONIS). First, the purpose of the document is provided. Second, the scope of the testing is identified. Third, the organization of the document is summarized.

B.4.1.1 Purpose of the Document

This document represents validation tests for a ONIS. Tester can use it to test the system and developers can use it to maintain the system.

B.4.1.2 Scope of the Testing

The scope of testing of ONIS is restricted to test the correctness, usability, tolerance and portability. Parser Module is the most important module in this system, so unit testing for Parser Module is also provided in this document. System testing verifies the correctness by comparing the solutions of ONIS with those of Matlab and Maple. ONIS will be run on MacOX and Windows operation system to test portability. Performance testing focuses on testing the speed of ONIS.

B.4.1.3 Organization of the Document

Section B.4.1 is an introduction to the report. Section B.4.2 shows the unit testing which focuses on the Parser Module. Two test cases are provided in this section to verify the correctness of the parser. Section B.4.3 lists test cases that are categorized by function types to test the correctness of ONIS. Also in this section, test cases, which test the tolerance, usability and performance of the ONIS, are provided. Two traceability matrixes are given in Section B.4.4 which shows the association between test cases and requirements, including functional and nonfunctional requirement, that are specified in the SRS as well as the relationship between test cases and leaf modules introduced in MG. Section B.4.5 presents the result and analysis.

B.4.2 Unit Testing

The unit testing shows the test cases for Parser Module which is used to parse the input function $f(x)$. The input of this module is the string of $f(x)$ and the output is the object of parse tree.

B.4.2.1 Test Case 1: Test Case for Unit Testing

Test case 1 is used to test the operator +, -, *, /, (, and).

- Input: (3 * x + 4) / (x - 2)

- OS: Mac OS X and Windows

143

- Expected Output:
  The output should be a parser tree like Figure B.3.

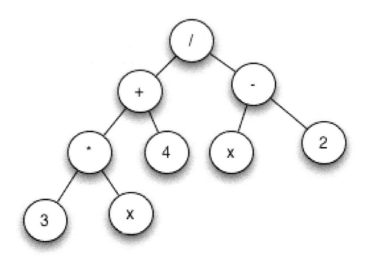

Figure B.3: Parse Tree of Test Case 1

- Test Result: PASSED

B.4.2.2   Test Case 2: Test Case for Unit Testing

Test case 2 is used to test the syntax of some programming language C build-in functions which are used in ONIS.

- Input: sin(x + log(x + 3)) / exp(2 * x) + pow(2, x)

- OS: Mac OS X and Windows

- Expected Output:
  The expected output should be a parser tree like Figure B.3.

- Test Result: PASSED

B.4.3   System Testing

Routines from a FORTRAN library, Quadpack, are used in ONIS to calculate integration according to different characteristics of the input function; therefore, system testing

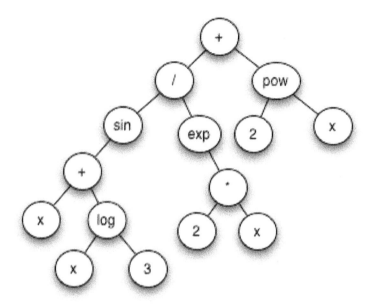

Figure B.4: Parse Tree of Test Case 2

focuses on testing those routines. System testing is categorized by function types. The expected outputs are the solutions provided by Matlab and Maple, which is used to validate the correctness of the solution of ONIS. Moreover, test cases for testing non-functional requirement such as performance, tolerance and usability are provided also.
The following is the table of function type.

- FT1: $f(x)$ is a normal function.

- FT2: $f(x)$ has end point singularities on $[a, b]$.

- FT3: $f(x)$ has an oscillatory behavior of nonspecific type and and no singularities.

B.4.3.1 Test Cases

Test Case 1: Test Case for FT1

- Input:

- $f(x) = x^2 + 3$
  - Input string of f(x) = x*x+3
  - a = -999,999, b = 999,999, epsrel = 0.00001

- OS: Unix and Windows

- Expected Output:

  - The solution of Matlab: 6.6666e+17

  - The solution of Maple: 6.666646e+17

- Test Result:
  result: 6.66665e+17
  absolute error: 0.00533636
  number of function evaluations: 21

- Conclusion: Test case 1 is used to test the normal function of $f(x)$ with a big interval [a, b]. The back end routine used in ONIS is QNG routine in Quadpack library. The solutions of ONIS, Matlab and Maple are almost the same, i.e., ONIS can handle this kind of problem.

Test Case 2: Test Case for FT2

- Input:

  - $$f(x) = \frac{1}{\sqrt{(1-x)} - 1}$$
  - Input string of f(x) = 1/(sqrt(1-x)-1)

  - a = 0.75, b = 1, epsrel = 0.00001

- OS: Unix and Windows

- Expected Output:

  - The solution of Matlab: -0.386294
  - The solution of Maple: -0.386294

146

- Test Result:
  result: -0.386294
  absolute error: 8.89672e-06
  number of function evaluations: 189

- Conclusion: Test case 2 is used to test $f(x)$ with end point singularities on $[a, b]$. The back end routine used in ONIS is QAGS routine in Quadpack library. The solutions of ONIS, Matlab and Maple are almost the same, i.e. ONIS can handle the problem FT2.

Test Case 3: Test Case for FT2

- Input:

    - $f(x) = \sqrt{x} * \log(x)$

    - Input string of f(x) = sqrt(x) * log(x)

    - a = 0, b = 999,999, epsabs = 0, epsrel = 0.00001

- OS: Unix and Windows

- Expected Output:

    - The solution of Matlab: 8.7659e+09
    - The solution of Maple: 8.76588e+09

- Test Result:
  result: 8.76588e+09
  absolute error: 7194.97
  number of function evaluations: 315

- Conclusion: The solutions of ONIS, Matlab and Maple are almost the same. Given a big interval [a, b], ONIS also can get the correct solution. So, ONIS can handle the problem FT2.

Test Case 4: Test Case for FT3

- Input:

- $f(x) = \cos(100\sin(x))$

- Input string of f(x) = cos(100*sin(x))

- a =0, b = 3.14, epsrel = 0.00001

- OS: Unix and Windows

- Expected Output:

    - The solution of Matlab: 0.0612

    - The solution of Maple: 0.0612015

- Test Result:
  result: 0.0612015
  absolute error: 7.43153e-09
  number of function evaluations: 427

- Conclusion: Test case 4 is used to test the $f(x)$ with an oscillatory behavior of nonspecific type and and no singularities. The back end routine used in ONIS is QAG routine with parameter KEY=6 in Quadpack library. The solution of ONIS, Maple and Matlab are almost same, when interval [a, b] is not too big.

Test Case 5: Test Case for FT3

- Input:

    - $f(x) = \cos(100\sin(x))$

    - Input string of f(x) = cos(100*sin(x))

    - a =-999999, b = 999999, epsrel = 0.01

- OS: Unix and Windows

- Expected Output:

    - The solution of Matlab: 6.8678e+05

148

– The solution of Maple: 686263

- Test Result:
  result: 40184.8
  absolute error: 1.16516e+06
  number of function evaluations: 121939
  Note: maximum number of function evaluations has been achieved.

- Conclusion: Test case 6 is also used to test the $f(x)$ with an oscillatory behavior of nonspecific type and and no singularities, which is almost the same as test case 6, but with a big interval [a,b]. The solution of ONIS is different, that means, ONIS cannot handle problem FT4 with big interval.

Test Case 6: Test Cases for Usability Testing
Test case 6 is used to test usability of ONIS. i.e. according to SRS, to test whether ONIS is easy to learn and use.

- Input:

  – $f(x) = x^2 + 3$

  – a = -999,999, b = 999,999, epsabs = 0, epsrel = 0.00001

- OS: Unix and Windows

- Output: 6.66665e+17

- Time: 1.5 minute

- Conclusion: After getting a basic knowledge of ONIS, a user who didn't use ONIS before spent about 1.5 minute finishing this simple test case and got the final solution. Although this user got some basic introduction before he uses ONIS, he still made some mistakes. So, if there is some documents, for example help files or user manuals, that users can follow, users will make less mistakes and it will be easy for users learn new application quickly

Test Case 7: Test Cases for Tolerance Testing

- Input:

  – $f(x) = x^2 + 3$

- Input string of f(x) = x*x+3

  - a = -999,999, b = 999,999, epsrel = 0.00001

- OS: Unix and Windows

- Expected Output:

  - The solution of Matlab: 6.6666e+17

  - The solution of ONIS: 6.66665e+17

  - $tolerance = \left| \frac{y_{calculated} - y_{true}}{y_{true}} \right| = \left| \frac{6.66665e+17 - 6.6666e+17}{6.6666e+17} \right| \leq 7.5e - 5$

- Conclusion: The tolerance reaches the demand for tolerance in SRS.

Test Case 8: Test Cases for Performance Testing

- Input:

  - $f(x) = x^2 + 3$

  - Input string of f(x) = x*x+3

  - a = -999,999, b = 999,999, epsrel = 0.00001

- OS: Unix and Windows

- Expected Output:

  - The solution of Matlab: 6.6666e+17
  - Response time of Matlab: 1.7300 second
  - The solution of ONIS: 6.66665e+17
  - Response time of ONIS: 0
  - Conclusion: For this test case, the response time of ONIS is 0, so the speed of ONIS is much faster than Matlab.

150

	1	2	3	4	5	6	7	8
IV	X	X	X	X	X	X	X	X
OV	X	X	X	X	X	X	X	X
N1								
N2								X
N3							X	
N4						X		
N5	X	X	X	X	X	X	X	X

Table B.12: Traceability Matrix for SRS

B.4.4 Traceability Matrix

B.4.4.1 Traceability Matrix for SRS

The traceability matrix defined in Table B.12 illustrates the relationship between requirements and test cases. The following is the functional and nonfunctional requirement table.

N1: Input Accuracy
N2: Performance
N3: Tolerance
N4: Usability
N5: Portability

According to SRS, all input data of the system are assumed to be accurate, so no test cases are put to test Input Accuracy.

B.4.4.2 Traceability Matrix for MG

Table B.13 is the traceability matrix shows the relationship between Module Guide and test cases. Mouse module is not used in ONIS; therefore, module M8 is not tested.

M1: Master Control Module
M2: Input Data Module
M3: Algorithm Module
M4: Parser Module
M5: Output Show Module
M6: Keyboard Input Module
M7: Screen Display Module
M8: Mouse Module

B.4.5 Results and Analysis

The testing results are showed in Section B.4.2 and B.4.3. The unit testing for Parser Module is tested. From the unit testing results, we know that the parser module works

	1	2	3	4	5	6	7	8
M1	X	X	X	X	X	X	X	X
M2	X	X	X	X	X	X	X	X
M3	X	X	X	X	X	X	X	X
M4	X	X	X	X	X	X	X	X
M5	X	X	X	X	X	X	X	X
M6	X	X	X	X	X	X	X	X
M7	X	X	X	X	X	X	X	X
M8								

Table B.13: Traceability Matrix for MG

well. The routines of Quadpack library: QNG, QAGS and QAG are tested. The solutions which get from these routines are almost the same as those of Maple and Matlab.

The portability of ONIS is good, ONIS can be run on MacOX and Windows operation system, but it needs two kinds of compiler, g++ and gfortran, to support it. Response time of ONIS is usually shorter than that of Maple and Matlab, so, in term of speed, the performance of ONIS is good. One test case is put to test tolerance, which reaches the requirement of SRS.

Appendix C

Unified Software Development Process

C.1 SRS for ONIS

This section provides the software requirement specification (SRS) for a one-dimensional numerical integration solver (ONIS).

C.1.1 Introduction

This section gives an overview of the Software Requirements Specification (SRS) for a One-dimensional Numerical Integration Solver (ONIS) using the Unified Software Development Process. The Unified Process uses the *Unified Modeling Language (UML)* when preparing all blueprints of the software system. The Unified Process is a use-case driven process (Jacobson et al., 1999, page 4).

A use case is a piece of functionality in the system that gives a user a result of value. Use cases capture functional requirements. All the use cases together make up the use-case model that describes the complete functionality of the system (Jacobson et al., 1999, Page 5). The phrase use case-driven refers to the fact that the project team uses the use cases to drive all development work, from initial gathering and negotiation of requirements through code.

C.1.1.1 Purpose of the Document

This SRS provides a description of a one-dimensional numerical integration of a function over a given interval. Most sections, except for Section 5 Functional Requirements, are taken from the SRS using the Parnas' Rational Design Process (PRDP) in Appendix B.1. To avoid duplications, but still present the complete version of this SRS, for sections with the same content, only the headings are provided in this document. The text for the corresponding section is not reproduced. The reader can refer to the corresponding parts in Appendix B.1.

The most important part of this document is Section C.1.4 Functional Requirement, as this section provides the use case model and a domain model of ONIS. The use case model defines the system's external behaviour and the requirements of the system, and therefore constrains the design and implementation. Hence, the use case model has a central role and is often said to drive the development process (Priestley, 2003, page 8). Based on the scope of ONIS, which is introduced in Section 1.3, a use case model, is provided in Section C.1.4. Normally, a use case model is supported by a domain model, which is a simple class diagram documenting the important business concepts and their relationships. The importance of the domain model is that it establishes the terminology that will be used for writing the descriptions of the use cases and offers some hope of removing ambiguity and lack of clarity in these descriptions. A domain model and use case descriptions are also provided in Section C.1.4 of this document.

C.1.1.2 Scope of the Software Product

Please see Appendix B.1.1.2.

C.1.1.3 Organization of the Document

This SRS follows the template given by (Smith, 2006). The rest of the document is organized as follows. Section C.1.2 provides the overall description of the system to make the requirements easier to understand. Section C.1.3 contains all the details of system requirements. Section C.1.4 introduces the non-functional requirements. Compared with Appendix B.1.1, Section C.1.5 describes the functional requirements of ONIS. It is a new section and does not exist in the SRS for PRDP. Section C.1.6 lists the solution validation strategies for this software. Other system issues, traceability matrix, list of possible changes in the requirements, and values of auxiliary constants are provided in the remainder of the sections.

C.1.2 General System Description

This section is the same as Appendix B.1.2. The system context is presented first. Then the characteristics of potential users are discussed. At the end of this section, some system constraints are described.

C.1.2.1 System Context

Please see Appendix B.1.2.1.

C.1.2.2 User Characteristics

Please see Appendix B.1.2.2.

C.1.2.3 System Constraints

Please see Appendix B.1.2.3.

C.1.3 Specific System Description

This section is almost the same as Appendix B.1.3. The only distinction is that the definitions of Use Case Model and Domain Model are added to Appendix B.1.3.2 Terminology Definition.

C.1.3.1 Background Overview

Please see Appendix B.1.3.1.

C.1.3.2 Terminology Definition

Use Case Model: a use case model describes (1) the system to be constructed, (2) the *actors* - representing a role played by a person or other entity that interacts with the system, and (3) the *use cases* - families of usage scenarios of the application, grouped into coherent cases of functionality (Lano, 2005, page 15).

Domain Model: domain model is a simple class diagram documenting the important business concepts and their relationships. It will be refined into a more comprehensive class diagram, which contains enough detail to form a basis for implementation (Priestley, 2003, page 49 - 51).

Other terminology definitions, such as Smooth and Singularity, are included in Appendix B.1.1.

155

C.1.3.3 Goal Statements

Please see Appendix B.1.3.3.

C.1.3.4 Theoretical Models

Please see Appendix B.1.3.4.

C.1.3.5 Data Definition

Please see Appendix B.1.3.5.

C.1.3.6 Assumption

Please see Appendix B.1.3.6.

C.1.3.7 Data Constraints

Please see Appendix B.1.3.7.

C.1.3.8 System Behaviour

Please see Appendix B.1.3.8.

C.1.4 Non-functional Requirements

Please see Appendix B.1.4.

C.1.5 Functional Requirements

Functional requirements capture the intended behavior of the system. Use cases have become a widespread practice for capturing functional requirements. This is especially true in the object-oriented community where they originated, but their applicability is not limited to object-oriented systems (Malan and Bredemeyer, 1999).

C.1.5.1 Use Case Model

The use case model presents a structured view of a system's functionality. It does this by defining a number of *actors*, which model the roles that users can play when interacting with the system, and describing the *use cases* that those actors can participate in. A *use case* describes a specific task that a user can achieve with the system. The *use case model* contains a set of *use cases*, which should define the complete functionality of the system. *Actors* and *use cases* of ONIS will be introduced in detail in the rest of this section.

Use Case Diagrams

A *use case diagram* summarizes in graphical form the different actors and use cases in a system, and shows which actors can participate in which use cases. Figure C.1 shows a use case diagram of ONIS. In ONIS, there is only one actor, *User*, who inputs data into ONIS, and one use case, which is *Calculate Integration*, showing scenarios of the use of ONIS.

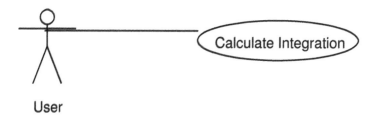

Figure C.1: Use Case Diagram

Use Case Descriptions
In this section, a use case template will be introduced. As well, descriptions for the use cases of ONIS will be provided.

1. Template of Descriptions

UML does not define a standard template for writing use case descriptions. The following template comes from (Priestley, 2003, page 348), which presents a representative list of headings under which a use case can be defined.

Name: The name of a use case, which is expressed as a short verb phrase, stating what task the user will accomplish with the use case.

Summary, or short description: It provides a one paragraph summary of what the use case accomplishes.

Actors: Lists the actors that are involved in the use case and the primary actor, who is responsible for initiating its execution.

Triggers: Triggers are the events that start the use case.

Preconditions: Preconditions summarize what must be true before the use case can start. Often preconditions state which other use cases must have run before the one being specified: a typical precondition might be 'the user has successfully logged on.'

Postconditions: Postconditions summarize what will be true when the use case has finished.

Courses of events, or scenarios: The basic, or normal, course of events should normally be presented as an unbroken sequence of interactions. The interactions within a course of events are normally numbered, for easy reference later on.

Alternative and exceptional courses of events: Alternative and exceptional courses of events can be written out in full. Often, however, it is adequate simply to specify the points at which the alternative flow diverges from the basic course of events.

This can be done by giving the number of the step where behaviour can vary and specifying what condition causes the divergence.

Extension points: This section should list the points in the course of events at which extensions can take place, and gives the condition or event that determines whether or not this will happen. The extensions themselves should be written as separate use cases; otherwise, an alternative course of events can be specified.

Inclusions: This section simply summarizes the use cases that are included into the use case being defined. The points at which inclusion takes place should be specified in the courses of events. This section is redundant if a use case diagram showing the same information is readily available.

2. Descriptions

The followings are the descriptions for the use case *Calculate Integration*, which is described using the template introduced in the previous section.

Name: Calculate Integration

Summary: The user enters into ONIS the following data: a lower bound, an upper bound, a function type, a function, function parameters sometimes, an expected absolute error and an expected relative error. ONIS checks the validity of the data as it is entered. Then, ONIS calculates the integration according to the user's input data. Finally, ONIS shows the results to the user, which includes an approximation to the integral, an estimate of the absolute error and the total number of function evaluations needed, or ONIS returns an error code.

Actors: *User.*

Triggers: The user runs ONIS.

Preconditions: The user has successfully started ONIS.

Postconditions: The system is succeeded in showing the results on the screen.

Courses of events:

1. The user enters a lower bound.

2. The system verifies the input lower bound. If the input data is correct, ONIS will go on and let the user input additional data. Otherwise, ONIS will stop and show an error message until the user enter the correct data.

3. The user enters an upper bound.

4. The system verifies the input upper bound. If the input data is correct, ONIS will go on and let the user input additional data. Otherwise, ONIS will stop and show an error message until the user inputs correct data.

5. The user enters an expected absolute error.

6. The system verifies the input expected absolute error. If the input data is correct, ONIS will go on and let the user input additional data. Otherwise, ONIS will stop and show an error message until the user inputs correct data.

7. The user enters an expected relative error.

8. The system verifies the input expected relative error. If the input data is correct, ONIS will go on and let the user input additional data. Otherwise, ONIS will stop and show an error message until the user inputs correct data.

9. The user enters a function.

10. The system verifies the input function. If the input data is correct, ONIS will go on and let the user input additional data. Otherwise, ONIS will stop and show an error message until the user inputs correct data.

11. The user enters a function type.

12. The system verifies the input function type. If the input data is correct, ONIS will go on and let the user input additional data. Otherwise, ONIS will stop and show an error message until the user inputs correct data.

13. The user enters function parameters, if the function type is CT_QAWO, CT_QAWS and CT_QAWC. The definition of the function type is in Section 3.5 Data Definition.

14. The system verifies the input function parameters. If the input data is correct, ONIS will go on and let the user input additional data. Otherwise, ONIS will stop and show an error message until the user input correct data.

15. The system returns the results.

16. The system shows an approximation to the integral, an estimate of the absolute error, and the total number of function evaluations needed to the *User*

Alternative and exceptional courses of events: The user stops inputting data and exits ONIS, then, the system is terminated.

The user input correct data to the system. The system selects a suitable algorithm routine to calculate the integration, but can not get an approximation to the integral and returns an error code.

Extension points: N/A

Inclusions: N/A

C.1.5.2 Domain Model

Use cases are intended to be comprehensible to both developers and users of the system. They are therefore described using terminology taken from the business domain rather than from implementation or computer-oriented vocabularies. A domain model, a class diagram showing the most important concepts in ONIS and the relationships between them, is carried out parallel with use case modeling to describe the business concepts that are used in the use case descriptions.

Diagram for Domain Model

Figure C.2 shows a domain model for ONIS. According to the user case model, four classes are defined for this system, which are *Master Control, Input Data, Calculating* and *Output Show.*

Glossaries

The following glossaries are a core vocabulary used in the domain model, which can help developers and users to talk about the system. Some vocabularies, such as *res* and *abserr*, are already defined in Appendix B.3.5 Data Definition but with different names. In this case, names in Definition are put right behind the corresponding vocabularies with bracket. For example, res (y) and abserr (ε_a)

MasterControl: A controller to control the sequences of the system.

InputData: Data which users input into ONIS.

Calculating: Calculating integration.

OutputShow: data which the system output to the user to show the calculation results.

a: Lower limit of integration.

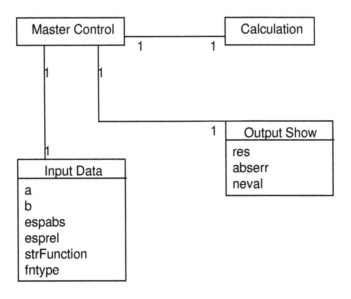

Figure C.2: Domain Model

b: Upper limit of integration.

epsabs: Absolute accuracy requested.

epsrel: Relative accuracy requested.

strFunction($f(x)$): Input function.

fntype(C): Characteristics of the input function.

res(y): Approximation to the integral.

abserr(ε_a): Estimate of the absolute error.

neval: Number of function evaluations.

Some vocabularies in the above glossaries are also in Appendix B.1 with the same meanings, but using the different names. Table C.1 presents the relationship between them.

Symbol in Data Constraints	Vocabulary in domain model
a	$Lowerbound$
b	$Upperbound$
$f(x)$	$Function$
C	$fntype$
$epsabs$	$Epsabs$
$epsrel$	$Epsrel$
y	$result$
ε_r	$abserr$
ε_a	$abserr$
$funcount$	$evalnum$

Table C.1: Table of Symbols

C.1.6 Solution Validation Strategies

This section is same as Appendix B.1.5.

C.1.6.1 Relative Comparison between Computed Solution and True Solution

Please see Appendix B.1.5.1.

C.1.6.2 Compare Solution with a Third-party System

Please see Appendix B.1.5.2.

C.1.6.3 Interval Arithmetic Method

Please see Appendix B.1.5.3.

C.1.6.4 Using Specific Test Cases

Please see Appendix B.1.5.4.

C.1.7 Other System Issues

This section is same as Appendix B.1.6.

C.1.7.1 Open Issues

Please see Appendix B.1.6.1.

C.1.7.2 Off-the-Shelf Solutions

Please see Appendix B.1.6.2.

C.1.7.3 Waiting Room

Please see Appendix B.1.6.3.

C.1.8   Traceability Matrix

Please see Appendix B.1.7.

C.1.9   Values of Auxiliary Constants

Please see Appendix B.1.8.

C.2 SDS for ONIS

This section provides the software design specification (SDS) for a one-dimensional numerical integration solver (ONIS).

C.2.1 Introduction

This section gives an overview of the Software Design Specification (SDS) for a One-dimensional Numerical Integration Solver (ONIS) using the unified Process. In terms of (Lano et al., 2002, page 2), software design is the organization of a software system into modules /subsystems /components/ classes or other units; the definition of behaviour and data storage responsibilities for these units, and the definition of interactions and collaborations between them which together meet the required functionalities of the system. In this document, first an analysis model and a design model will be introduced to realize the use-case model from SRS. Then, classes of ONIS will be provided to detail the information of ONIS. Also, to present the interactions and collaborations between classes, sequence diagrams and collaboration diagrams will be provided in this document.

There are 11 sections in this document. Section C.2.2 and Section C.2.3 provide anticipated changes and unlikely changes of ONIS. These two sections are same as Section B.2.3 and Section B.2.4 in Appendix B.2. Section C.2.4 contains the terminology definitions that will be used in this document. Section C.2.5 introduces the connections between the requirement and design. In Section C.2.6 and Section C.2.7, an analysis model and a design model are provided to present how to realize the use cases in SRS and how to decompose the system into classes. Section C.2.8 presents all the detail information of each class in ONIS. Section C.2.9 shows interaction diagrams that model the behaviour of ONIS. Section C.2.10 provides a traceability matrix to trace the relationships between the requirement and design. The last section, Section C.2.11, introduces the system architecture.

C.2.2 Anticipated Changes

Please see Appendix B.2.3.

C.2.3 Unlikely Changes

Please see Appendix B.2.4.

C.2.4 Terminology Definition

Analysis Model: An object model with the following purpose (1) to describe the requirements precisely; (2) to structure them in a way that facilitates understanding them, preparing them, changing them, and, in general, maintaining them; and (3) to work as an essential input for shaping the system in design and implementation (Jacobson et al., 1999, page 436).

163

Design Model: An object model that describes the physical realization of use cases and focuses on how functional and nonfunctional requirements together with other constraints related to the implementation environment impacts the system under consideration (Jacobson et al., 1999, page 436).

Analysis Class: An analysis class represents an abstraction of one or several classes and/or subsystems in the system's design. This abstraction has the following characteristics: (1) An analysis class focuses on handling functional requirements and postpones the handling of nonfunctional requirements. (2) An analysis class seldom defines or provides any interface in terms of operations. Instead, its behavior is defined by responsibilities on a higher, less formal level. (3) An analysis class defines attributes, although those attributes are also on a fairly high level. (4) An analysis class is involved in relationships, although those relationships are more conceptual than their design and implementation counterparts. (5) Analysis classes always fit one of three basic stereotypes: boundary, control, or entity (Jacobson et al., 1999, page 181,182).

Boundary Classes: Boundary Classes are in general used to model interaction between the system and its actors (i.e., users and external systems) (Jacobson et al., 1999, page 44).

Control Classes: Control Classes in the Unified Process are concerned more with controlling the interactions involved in a use case at the application level; they do not handle input and output (Priestley, 2003, page 82).

Entity Classes: Entity Classes are responsible for maintaining data (Priestley, 2003, page 83).

Design Class: A design class is a seamless abstraction of a class or similar construct in the system's implementation. This abstraction is seamless in the following sense: (1) The language used to specify a design class is the same as the programming language. Consequently, operations, parameters, attributes, types, and so on are specified using chosen programming language syntax. (2) The visibility of attributes and operations of a design class is often specified. (3) The relationships in which a design class is involved with other classes often has a straightforward meaning when the class is implemented. (4) The methods of a design class have straightforward mappings to the corresponding methods in the implementation of the class (Jacobson et al., 1999, page 217-219).

OCL: The Object Constraint Language (OCL) is a formal language to express side effect-free constraints. Users of the Unified Modeling Language can use OCL to specify constraints and other expressions attached to their models (IBM, 1997, page 1).

Attributes: An attribute is a description of a data field that is maintained by each instance of a class. Attributes must be named. In addition to the name, other pieces of information can be supplied, such as the type of the data described by the attribute or a default initial value for the attribute (Priestley, 2003, page 146).

Operations: Operations define the behaviour of instances of the class (Priestley, 2003, page 17).

164

Class Invariant: A class invariant is a property of a class that is intended to be true at all times for all instances of the class. However, the term 'invariant' is commonly used to refer only to constraints that restrict the possible values of a class's attributes (Priestley, 2003, page 263). The latter meaning of Class Invariant is used in this document.

Precondition and Postcondition: Defining an invariant for a class provides no guarantee that the operations of the class will ensure that the invariant is maintained. Preconditions and postconditions are special constraints that can be written for operations. As the names suggest, a precondition is something that must be true just before an operation is called and a postcondition is something that must be true after the operation has completed. These constraints should be written in such a way that if they are both true at the appropriate time, the invariant of the class will still be true when the operation has completed (Priestley, 2003, page 264).

Exceptions: An exception is an event, which occurs during the execution of a program, that disrupts the normal flow of the program's instructions (Sun Microsystems, 2008, 2008).

C.2.5 Connection Between Requirement and Design

The design of the system should satisfy the SRS. In this stage, an analysis model and a design model will be provided separately according to the use-case model in the SRS. As a use-case driven process, the unified process starts from the use case design. After use cases are decided, use cases will be realized to use-case realizations which are presented by analysis models and design models. Analysis Model and Design Model can also help decompose the whole system into different classes. The following diagram, Figure C.3, shows the traceability between models.

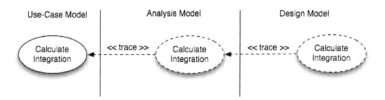

Figure C.3: Use-Case Realizations in the Analysis and Design Models

Also, a traceability matrix in Section 10 shows the connection between requirements and classes.

C.2.6 Analysis Model

Analysis model can help to achieve a more precise understanding of the requirements and it is described using the language of the developers, and can thereby introduce more formalism and be used to reason about the internal working of the system. Analysis model

Use-Case Model	Analysis Model
Described using the language of the customer	Described using the language of the developer
External view of the system	Internal view of the system
Structured by use cases; gives structure to the external view	Structured by stereotypical classes and packages; gives structure to the internal view
Used primarily as a contract between the customer and the developers on what the system should and should not to	Used primarily by developers to understand how the system should be shaped, i.e., designed and implemented
May contain redundancies, inconsistencies, etc. among requirements	Should not contain redundancies, inconsistencies, etc., among requirements
Captures the functionality of the system, including architecturally significant functionality	Outlines how to realize the functionality within the system, including architecturally significant functionality; work as a first cut at design
Defines use cases that are further analyzed in the analysis model	Defines use-case realizations, each one representing the analysis of a use case from the use-case model

Table C.2: Brief Comparison of the Use-Case Model and the Analysis Model

provides an overview of the system that may be harder to get by studying the results of design or implementation since too many details are introduced. Such an overview can be very valuable to newcomers to the system or to developers who maintain the system in general (Jacobson et al., 1999, page 178).

In this section, an analysis model, which is transformed from Use-case model in SRS, is provided. To clarify the relationships between the use-case model and analysis model, Table C.2 (Jacobson et al., 1999, page 175), a brief comparison of the use-case model and analysis model is provided as follows.

C.2.6.1   Analysis Model Diagram

Within the analysis model, use cases are realized by analysis classes and their objects. This is represented by collaborations within the analysis model. Figure C.4 below describes how the *Calculate Integration* use case is realized by a collaboration with a ≪trace≫ dependency between them, and that four classes participate and play roles in this analysis model. In this analysis model, the Solver Interface is a boundary classes, the Calculation is a control class, and the Algorithm and Parser are entity classes.

166

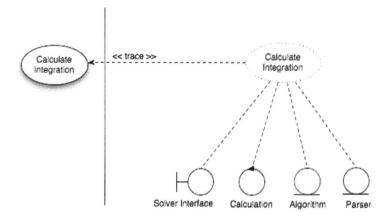

Figure C.4: Analysis Classes that Participate in a Realization of the Calculate Integration Use Case

C.2.6.2 Analysis Model Description Using a Collaboration Diagram

The sequence of actions in a use case begins when an actor invokes the use case by sending some form of message to the system. When considering the "inside" of the system, a boundary object will receive this message from the actor, i.e., Solver Interface object receives the input data from the user. The boundary object then sends a message to some other object, and so the involved objects interact to realize the use case, i.e., Solver Interface sends a message to a control object, Calculation, which controls the sequence of ONIS. Calculation calls the Algorithm object to select a suitable routine to calculate the integration. Figure is a collaboration diagram that describes how the *Calculate Integration* use-case realization is performed by a society of analysis objects. The diagram shows how the focus moves from object to object as the use case is performed and messages that are sent between the objects.

The following shows the flow of events of the diagram.

1. The User inputs all data needed to Solver Interface object.

2. The Solver Interface object verifies the input data and asks a Calculation controller object to perform the transition.

3. Calculation controller object asks Algorithm object to select a suitable algorithm routine to calculate the integration.

4. Algorithm object calls Parser object to parse the input function and get the values of function evaluations. Then Algorithm object gets the final results.

167

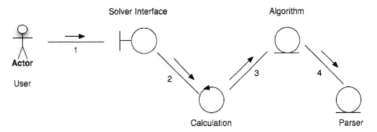

Figure C.5: A Collaboration Diagram for a Realization of the Calculate Integration Use Case

C.2.7 Design Model

The design model is an object that describes the physical realization of use cases by focusing on how functional and nonfunctional requirements, together with other constraints related to the implementation environment, impact the system under consideration. In addition, the design model serves as an abstraction of the system's implementation and is thereby used as an essential input to activities in implementation. Within the design model, use cases are realized by design classes and their objects. The following Table C.3 (Jacobson et al., 1999, page 219) describes the relationships between the analysis model and design model.

In Figure C.6, four analysis classes Solver Interface, Calculation, Algorithm and Parser participate in realizing the *Calculate Integration* use case in the analysis model. Also, in the design model, five design classes, Input Data, Output Show, Master Control, Algorithm and Parser, are refined from analysis classes, adapting to the implementation environment. Input Data and Output Show come from the boundary class Solver Interface, which control the interaction between ONIS and the user, i.e. Input Data helps the user input data to system and Output Show helps the system show the final calculating results to the user. MasterControl comes from the control class, Calculation, which controls the sequence of the system. Algorithm and Parser come from the same name the entity classes in the analysis model. The functionality of Algorithm is choosing a suitable routine to calculate the integration. Meanwhile, Parser helps to parse the input function and conduct function evaluations.

C.2.8 Class Description

In terms of the design model, classes that are defined in the design model are introduced in this section. First, the UML constraint language, OCL (Object Constraint Language),

168

Analysis Model	Design Model
Conceptual model, because it is an abstraction of the system and avoids implementation issues	Physical model, because it is a blueprint of the implementation
Design-generic (applicable to several designs)	Not generic, but specific for an implementation
Three (Conceptual) stereotypes on classes: ≪control≫, ≪entity≫, and ≪boundary≫	Any number of (physical) stereotypes on classes, depending on implementation language
Less formal	More formal
Less expensive to develop	More expensive to develop
Dynamic (but not much focus on sequence)	Dynamic (much focus on sequence)
Outlines the design of the system, including its architecture	Manifests the design of the system, including its architecture (one of its views)

Table C.3: Brief Comparison of the Analysis Model and the Design Model

which is used to define the class invariants, preconditions, postconditions and exceptions is described. Then, a brief system class diagram is showed to outline the structure of the whole system. Then, Fortran routines which are used in ONIS are introduced. Last, detailed information on each class is presented.

C.2.8.1 OCL Introduction

OCL is a means to express more complex properties of diagram elements, and interrelationships between elements. The grammars of OCL to introduce class invariants and operations are provided in this section. In addition, OCL basic data types and enumeration type are also presented in this section.

Class Invariant

As mentioned in Section 4 Terminology Definition, the term 'class invariant' in this document refers to constraints that restrict the possible values of a *class*'s attributes. OCL expression to introduce class invariants is as follows:

context TypeName **inv**: *Boolean expression*

The context keyword introduces the *context* for the expression. The keyword *inv* denotes

the stereotypes <<invariant>> of the constraint. The actual OCL expression comes after the colon. The boldface has no formal meaning, but is used to make the expressions more readable in this document. Note that all OCL expressions that express invariants are of the type Boolean (OMG, 2003, page 6).

Pre- and Postconditions The OCL expression can be part of a precondition or postcondition, corresponding to <<precondition>> and <<postcondition>> stereotypes of constraint associated with an operation or other behaviour feature. The context decla-

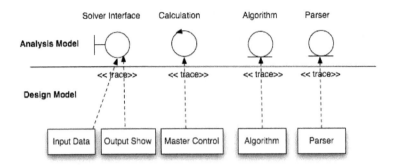

Figure C.6: Design Classes in the Design Model Tracing to Analysis Classes in the Analysis Model

ration in OCL uses the *context* keyword, followed by the type and operation declaration. The stereotype of constraint is shown by putting the labels 'pre:' and 'post:' before the actual preconditions and postconditions.

context Typename::operationName(param1 : Type1, ...): Return Type

pre: param1 > ...
post: result = ...
The reserved word *result* denotes the result of the operation, if there is one. The names

of the parameters (*param1*) can also be used in the OCL expression (OMG, 2003, page 8).

Reliability is a major characteristic of high-quality software. Software should behave well in nearly every situation. During the development process, errors as possible should be avoided, detected and removed; therefor, exception handling is an important part of ONIS. In this document, exceptions are described including in the postconditions using *If Expressions* in OCL. An *If Expression* results in one of two alternative expressions depending on the evaluated value of a condition (OMG, 2003, page 46). The detail information of exception handling is provided in Section 10 Exception Handling.

Data Types
In OCL, a number of basic types are predefined and available to the modeler at all time. These predefined value types are independent of any object model and part of the definition of OCL. The most basic value in OCL is a value of one of the basic type. There are four basic types in OCL which are Boolean, Integer, Real and String (OMG, 2003, page 10).

Enumeration Types
Enumerations are data types in UML and have a name, just like any other classifier. An enumeration defines a number of enumeration literals, that are possible values of enumeration. Within OCL one can refer to the value of an enumeration (OMG, 2003, page 11). For example, when we have data type named Gender with values 'female' or 'male' they can be used as follows:

context Person **inv**: gender = Gender::male

In ONIS, *Ctype, Ecodetype, TFLAG* and *SYMBOL* are defined as enumeration types. Figure 6, 7 and 8 below present their diagrams respectively.

C.2.8.2 System Class Diagram

Figure C.7 illustrates a brief class diagram of ONIS, showing the relationships among classes. There are 6 classes in this system, which are *MasterControl Class, InputData Class, Algorithm Class, Parser Class, Expression Class* and *OutputShow Class*. The link between classes is named *association*. The existence of an association between two classes indicates that instances of the classes can be linked at run-time. The arrowheads on the associations means the association can only be navigable in one direction. The symbol "1" and "1..*" at the association ends are *multiplicity* which means "exactly one object." and "at least one object", respectively. On this diagram, for example the multiplicities at the InputData and MasterControl end are all "1" which specifies that a MasterControl object can only be linked to one InputData instance.

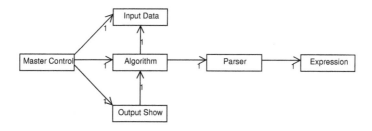

Figure C.7: A Brief Class Diagram of ONIS

C.2.8.3 Fortran Routines Introduction

Fortran Routines, QAWO, QAWS, QAWC, QNG, QAGS and QAG Quadpack library will be used in ONIS to calculate the integration according to the characteristics of an input function. To specify the information of the classes in ONIS clearly, as well as, let software developers better understand the functionalities of MasterControl and Algorithm class, the Fortran routines are introduced first. These Fortran routines are encapsulated in the operations in Algorithm class using the same name but with a prefix 'd' before the corresponding routine name. For example, QAWO will be encapsulated in dqawo() in Algorithm class. The meanings of these routines are provided as follows (FSU, 2008):

- If you can factor the integrand as $f(x)=w(x) \times g(x)$, where g is smooth on [a,b] and $w(x)=\cos(\omega \times x)$ or $\sin(\omega \times x)$ then use QAWO.

171

- Otherwise, if you can factor $f(x)=w(x) \times g(x)$ where g is smooth and $w(x)=(x - a)^{\alpha} \times (b - x)^{\beta} \times (log(x - a))^l \times (log(b - x))^k$ with k, $l = 0$ or 1, and *alpha*, *beta* greater than -1, then use QAWS.

- Otherwise, if you can factor $f(x)=w(x) \times g(x)$ where g is smooth and $w(x) = 1/(x - c)$ for some constant c, use QAWC.

- Otherwise, if the integrand is smooth, use QNG.

- Otherwise, if the integrand has end point singularities, use QAGS.

- Otherwise, if the integrand has an oscillatory behavior of nonspecific type, and no singularities, use QAG with KEY=6.

C.2.8.4   MasterControl Class

MasterControl Class includes a main() function, controlling the execution sequence of the ONIS. The sequence is as follows:

1. It calls the operations of InputData class, which are *setLowerbound*, *setUpperbound*, *setEpsabs*, *setEpsrel*, *setFunction* and *setFntype* to get users' input data, which are lower limit of integration, upper limit of integration, a requested absolute accuracy, a requested relative accuracy, an input function and characteristics of the input function respectively. The above operations will be introduced in detail in Section 8.4 InputData class.

2. After that, MasterControl call one Algorithm operation, *dqawo*, *dqaws*, *dqawc*, *dqng*, *dqags* or *dqag* according to the characteristics of users' input function *fntype*. The mapping between characteristics of the input function and the Algorithm operation is described following:
   if *fntype* = Ctype::CT_QAWO then *dqawo*

   else if *fntype* = Ctype::CT_QAWS then *dqaws*
   else if *fntype* = Ctype::CT_QAWC then *dqawc*
   else if *fntype* = Ctype::CT_QNG then *dqng*
   else if *fntype* = Ctype::CT_QAGS then *dqags*
   else if *fntype* = Ctype::CT_QAG then *dqag*
   endif
   [**Note:**] *fntype* : *Ctype* is an enumeration type. Figure C.8 shows the diagram of

   Ctype with its literals. The literals of this type are defined according to characteristics of input function which is in Section C.1.3.5 Data Definition.

   When the *fntype* is Ctype::CT_QAWO, Ctype::CT_QAWS or Ctype::CT_QAWC, additional parameters are needed for the corresponding operations in Algorithm class, which are described in the following:

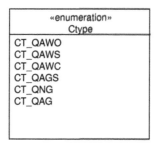

«enumeration» Ctype
CT_QAWO CT_QAWS CT_QAWC CT_QAGS CT_QNG CT_QAG

Figure C.8: Enumeration Type - Ctype

- If *fntype* = Ctype::CT_QAWO, before *dqawo* is invoked, operations in Input-Data class, *setOmega* and *setIntegr*, are executed to set parameters *omega* and *integr* which are input parameters of *dqawo*. Note *integr* can only be set to 1 when $w(x) = \cos(\omega * x)$ or 2 when $w(x) = \sin(\omega * x)$.

- If *fntype* = Ctype::CT_QAWS, before *dqaws* is invoked, operations in Input-Data class, *setAlphaBeta* and *setIntegr*, are executed to set parameters *alpha*, *beta* and *integr*, which are input parameters of *dqaws*. *integr* can only be set to 1, 2, 3 or 4. The detail information is as follows:
  1 when $w(x) = (x-a)^{\alpha} \times (b-x)^{\beta}$ or
  2 when $w(x) = (x-a)^{\alpha} \times (b-x)^{\beta} \times log(x-a)$ or
  3 when $w(x) = (x-a)^{\alpha} \times (b-x)^{\beta} \times log(b-x)$ or
  4 when $w(x) = (x-a)^{\alpha} \times (b-x)^{\beta} \times log(x-a) \times log(b-x)$.

- If *fntype* = Ctype::CT_QAWC, before *dqawc* is invoked, an operation in InputData class, *setC*, is executed to set the parameter *c*, which is the input parameter of *dqawc*.

3. After Algorithm Class calculates the integration and has the final solution, Master-Control calls OutputShow Class to show the results, which are got from Algorithm class, on the screen.

Sequence diagram which is defined in UML is a good way to specify the integrations between objects as well as system behaviours. The above sequence is illustrated using *sequence diagram* in Figure 11 in Section 9.2.

Attributes
None.

Class Invariant
None.

Operations

There is only one operation, main(), in MasterControl Class. The following is the description of main(), which includes the pre- and postcondition of main() presented by OCL expressions. In addition, descriptions and assumptions are also provided using natural languages.

context MasterControl :: main()
pre: true
post: true

Assumptions: main() is executed before any other operations in ONIS. Moreover, during the running time, the instances of InputData, Algorithm and OutputShow class are linked only once with the instance of MasterControl class.
Description: main() is the point where ONIS starts its execution. As well, it controls the executive sequence of ONIS. The sequence of main is presented in Figure 11 in Section 9.2.

C.2.8.5 InputData Class

InputData Class is used to help the user input data, which includes a function, characteristics of the input function, an upper bound, a lower bound, an absolute accuracy requested and a relative accuracy requested, to ONIS. In some of input functions, for instance *fntype* equals to Ctype::QAWO, Ctype::QAWS or Ctype::QAWC, as mentioned in MasterControl class, additional parameters such as *omega*, *alpha*, *beta* and *c* are needed.

Attributes

Table C.4 shows the attributes of Input Data Class, which includes the name, data type and descriptions of each attribute in this class.

Class Invariant
context InputData
inv: MIN_A $< a <$ MAX_A
inv: MIN_B $< b <$ MAX_B

Attributes Initialization

To satisfy the class invariants, initial values are set to some attributes in the class using the following syntax in OCL to indicate the initial value of the attributes (OMG, 2003, page 10).

context Typename :: attributeName: Type
init: – some expression resp renting the initial value

Name	Data Type	Descriptions
a	*Real*	Lower limit of integration
b	*Real*	Upper limit of integration
epsabs	*Real*	Absolute accuracy requested
epsrel	*Real*	Relative accuracy requested
fntype	*Ctype*	Characteristics of the input function
strFunction	*String*	Input function
c	*Real*	a constant in the weight function $w(x) = 1 / (x\text{-}c)$ when *fntype* = "Ctype::CT_QAWC"
omega	*Real*	a parameter in the weight function when *fntype* = "Ctype::CT_QAWO"
alpha	*Real*	a parameter in the weight function when *fntype* = "Ctype::CT_QAWS"
beta	*Real*	a parameter in the weight function when *fntype* = "Ctype::CT_QAWS"
integr	*Integer*	a parameter to show the type of the weight function when *fntype* = "Ctype::CT_QAWS" or "Ctype::CT_QAWO"

Table C.4: Attributes of Input Data Class

context InputData :: *a* : *Real*
init: *a* = MIN_A + 1

context InputData :: *b* : *Real*
init: *b* = MAX_B - 1

Operations
In this section, operations in InputData class are introduced. The values of MIN_A, MAX_A, MIN_B, MAX_B, MAX_EPSABS and MAX_EPSREL which are existed in the following operations can be referred to Section C.1.9 Values of Auxiliary Constants.

context InputData :: setLowerbound()
pre: true
post: if *a1* >= MIN_A and *a1* <= MAX_A
        then *a* = *a1*
        else ExceptionID = LowerBound_input_invalid
        endif
Assumptions: setLowerbound() is invoked before setUpperbound()
Description: setLowerbound() receives a real type lower bound value *a1* from the keyboard and stores this value to the attribute *a*. If *a1* is not satisfied the condition MIN_A< *a1* <MAX_A, the system will jump to exception handler, which will catch the ExceptionID and show the corresponding error message on the screen.

175

[**Note:**] The exception handling methods of the other operations in InputData class are all very similar as setLowerbound(). The only differences are their ExceptionID and corresponding error messages. To simplify the design document, in the following descriptions of the operations, the specification of exception handling is not included. In stead, Section 10 Exception Handling will present the exception handling strategies of ONIS in detail.

context InputData :: setUpperbound()
pre: true
post: if $b1 > \text{MIN\_B}$ and $b1 < \text{MAX\_B}$ and $b1 >= self.a$
        then $b = b1$
        else ExceptionID = UpperBound_input_invalid
        endif
Assumptions: setUpperbound() is invoked after setLowerbound(), because after the user inputs an upper bound $b1$, the program needs to compare the value of $b1$ and a to satisfy the condition $b1 \geq a$.
Description: setUpperbound() receives a real type upper bound value $b1$ from the keyboard and stores this value to the attribute b

[**Note:**] In OCL, the contextual instance *self* is of the type which owns the operation as a feature (IBM, 1997, Page 4). the value of a property on an object that is specified by a dot followed by the name of the property. For instance, *self.a* is the value of the property a on *self*.

context InputData :: setEpsabs()
pre: true
post: if $0 < epsabs1 < \text{MAX\_EPSABS}$
        then $epsabs = epsabs1$
        else ExceptionID = Epsabs_input_invalid
        endif
Assumptions: None
Description: setEpsabs() receives a real type absolute accuracy requested value *epsabs1* from the keyboard and stores this value to the attribute *epsabs*

context InputData :: setEpsrel()
pre: true
post: if $0 < epsrel1 < \text{MAX\_EPSREL}$
        then $epsrel = epsrel1$
        else ExceptionID = Epsrel_input_invalid
        endif
Assumptions: None
Description: setEpsrel() receives a real type relative accuracy requested value *epsrel1* from the keyboard and stores this value to the attribute *epsrel*

context InputData :: setFunction()
pre: true
post: if *strFunction1.size* $<> 0$
           then *strFunction = strFunction1*
           else ExceptionID = Function_input_invalid
           endif
Assumptions: None
Description: setFunction() receives a function string *strFunction*1 from the keyboard and stores this string to *strFunction*

context InputData :: setC()
pre: *self.fntype* = Ctype::CT_QAWC
post: if *self.fntype* = Ctype::CT_QAWC
          then if $c1 > self.a$ and $c1 < self.b$
             then $c = c1$
             else ExceptionID = C_input_invalid
             endif
          else ExceptionID = Fntype_input_invalid
          endif
Assumptions: setC() should be executed after setLowerbound() and setUpperbound(), because *c1* should satisfy the condition $c > self.a$ and $c < self.b$
Description: setC() receives a Real type value *c1* from the keyboard and stores this value to *c*

context InputData :: setIntegr()
pre: *self.fntype* = Ctype::CT_QAWO or Ctype::CT_QAWS
post: if *self.fntype* = Ctype::CT_QAWO or *self.fntype* = Ctype::CT_QAWS
          then *integr = integr1*
          else ExceptionID = Integr_input_invalid
          endif
noindent Assumptions: None
Description: setIntegr() receives a Integer type value *integr1* from the keyboard and stores this value to *integr*

context InputData :: setOmega()
pre: *fntype* = Ctype::CT_QAWO
post: if *self.fntype* = Ctype::CT_QAWO
          then *omega = omega1*
          else ExceptionID = Fntype_input_invalid
          endif
Assumptions: None
Description: setOmega() receives a Real type value *omega1* from the keyboard and stores this value to *omega*

177

context InputData :: setAlphaBeta()
pre: *fntype* = Ctype::CT_QAWS
post: if *self.fntype* = Ctype::CT_QAWS
                            then if $alpha > -1$ and $beta > -1$
                                     then $alpha = alpha1$; $beta = beta1$
                                     else ExceptionID = Omega_input_invalid
                                     endif
                          else ExceptionID = Fntype_invalid
                          endif
Assumptions: None
Description: setAlphaBeta() receives Real type value *alpha1* and *beta1* from the keyboard and stores them to *alpha* and *beta* respectively

context InputData :: getLowerbound() : Real
pre: true
post: result = *self.a*
Assumptions: getLowerbound() is executed after setLowerbound()
Description: Get the value of the attribute *a*

context InputData :: getUpperbound() : Real
pre: true
post: result = *self.b*
Assumptions: getUpperbound() is executed after setUpperbound()
Description: Get the value of the attribute *b*

context InputData :: getEpsabs() : Real
pre: true
post: result = *self.epsabs*
Assumptions: getEpsabs() is executed after setEpsabs()
Description: Get the value of the attribute *epsabs*

context InputData :: getEpsrel() : Real
pre: true
post: result = *self.epsrel*
Assumptions: getEpsrel() is executed after setEpsrel()
Description: Get the value of the attribute *epsrel*

context InputData :: getFntype() : Real
pre: true
post: result = *self.fntype*
Assumptions: getFntype() is executed after setFntype()
Description: Get the value of the attribute *fntype*

context InputData :: getFunction() : String
pre: true
post: result = *self.strFunction*
Assumptions: getFunction() is executed after setFunction()
Description: Get the value of the attribute *strFunction*

context InputData :: getC() : Real
pre: true
post: result = *self.c*
Assumptions: getC() is executed after setC()
Description: Get the value of the attribute *C*

context InputData :: getOmega() : Real
pre: true
post: result = *self.omega*
Assumptions: getOmega() is executed after setOmega()
Description: Get the value of the attribute *omega*

context InputData :: getAlpha() : Real
pre: true
post: result = *self.alpha*
Assumptions: getAlpha() is executed after setAlfaBeta()
Description: Get the value of the attribute *alpha*

context InputData :: getBeta() : Real
pre: true
post: result = *self.beta*
Assumptions: getBeta() is executed after setAlfaBeta()
Description: Get the value of the attribute *beta*

context InputData :: getIntegr() : Integer
pre: true
post: result = *self.integr*
Assumptions: getIntegr() is executed after setIntegr()
Description: Get the value of the attribute *Integer*

C.2.8.6   Algorithm Class

Algorithm Class is in charge of calculating the integration. As mentioned before, Fortran
routines, QAWO, QAWS, QAWC, QNG, QAGS, QAG in Quadpack library will be encapsulated in this class to help calculate the integration according to the characteristics

179

Name	Data Type	Descriptions
res	*Real*	The estimated value of the integral
abserr	*Real*	The estimated of absolute error
neval	*Integer*	The number of times the integral was evaluated
ier	*Ecodetype*	Error code.

Table C.5: Attributes of Algorithm Class

of the input function.

Attributes
Table C.5 shows the attributes of Algorithm Class.

[Note:] *ier: Ecodetype* is an enumeration type. Figure 7 shows the diagram of *Ecodetype* with its literals. The literals of this type are defined according to Error Code Type which is in Section 3.5 Data Definition in Section C.1.

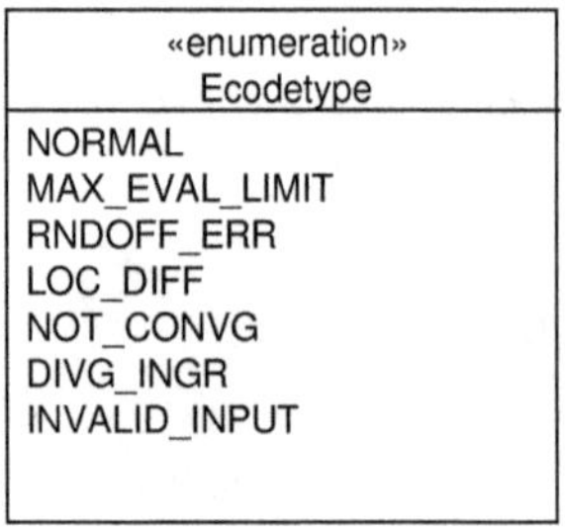

```
┌─────────────────────────┐
│      «enumeration»       │
│       Ecodetype          │
├─────────────────────────┤
│  NORMAL                  │
│  MAX_EVAL_LIMIT          │
│  RNDOFF_ERR              │
│  LOC_DIFF                │
│  NOT_CONVG               │
│  DIVG_INGR               │
│  INVALID_INPUT           │
│                          │
└─────────────────────────┘
```

Figure C.9: Enumeration Type - Ecodetype

Class Invariant
None.
Operations

context Algorithm :: dqawo(*sfun: String, a: Real, b: Real, omega: Real, integr: Integer, epsabs: Real, epsrel: Real*)
pre: *self.InputData.fntype* = Ctype::CT_QAWO
post: if *self.InputData.fntype* = Ctype::CT_QAWO
              then if *self.InputData.integr* = 1 or *self.InputData.integr* = 2
                then result = Tuple {

$$self.res = \text{approximate value of } y = \int_a^b f(x)\,dx,$$

180

$$self.abserr = \text{approx}|\, y - res\,|,$$
$$self.neval = \text{number of integrand evaluations},$$
{Ecodetype::NORMAL, Ecodetype::MAX_EVAL_LIMIT,
Ecodetype::RNDOFF_ERR, Ecodetype::LOC_DIFF, Ecodetype::NOT_CONVG, Ecodetype::DIVG_INGR} -> includes *self.ier*
}
else *self.ier* = Ecodetype::INVALID_INPUT
endif
else ExceptionID = Fntype_invalid
endif

Note: The meanings of literals of Ecodetype can be referred to Values of Auxiliary Constants in Section C.1

Assumptions: None

Descriptions: dqawo() calculates the integration of the input function whose *Ctype* is "CT_QAWO." dqawo uses a Fortran routine qawo.f (Netlib, 2008d, 2008) which is in Quadpack library to calculate the integration.

context Algorithm :: dqaws(*sfun: String, a: Real, b: Real, alpha: Real, beta: Real, integr: Integer, epsabs: Real, epsrel: Real*)

pre: *self.InputData.fntype* = Ctype::CT_QAWS

post: if *self.InputData.fntype* = Ctype::CT_QAWS
then if {1..4} -> includes(*self.InputData.integr*)
then result = Tuple {

$$self.res = \text{approximate value of } y = \int_a^b f(x)\,dx,$$
$$self.abserr = \text{approx}|\, y - res\,|,$$
$$self.neval = \text{number of integrand evaluations},$$
{Ecodetype::NORMAL, Ecodetype::MAX_EVAL_LIMIT,
Ecodetype::RNDOFF_ERR, Ecodetype::LOC_DIFF} -> includes *self.ier*
}
else *self.ier* = Ecodetype::INVALID_INPUT
endif
else ExceptionID = Fntype_invalid
endif

Assumptions: None

Descriptions: dqaws() calculates the integration of the input function whose *Ctype* is "CT_QAWS." dqaws() uses a Fortran routine qaws.f (Netlib, 2008e, 2008) which is in Quadpack library to calculate the integration.

context Algorithm :: dqawc(*sfun: String, a: Real, b: Real, c: Real, epsabs: Real, epsrel: Real*)

pre: *self.InputData.fntype* = Ctype::CT_QAWC

post: if *self.InputData.fntype* = Ctype::CT_QAWC
then if $a < c$ and $c < b$
then result = Tuple {

$$self.res = \text{approximate value of } y = \int_a^b f(x)\, dx,$$

$$self.abserr = \text{approx}|\, y - res\, |,$$

$$self.neval = \text{number of integrand evaluations},$$

{Ecodetype::NORMAL, Ecodetype::MAX_EVAL_LIMIT,
Ecodetype::RNDOFF_ERR, Ecodetype::LOC_DIFF} -> includes *self.ier*
}
else *self.ier* = Ecodetype::INVALID_INPUT
endif
else ExceptionID = Fntype_invalid
endif

Assumptions: None

Descriptions: dqawc() calculates the integration of the input function whose *Ctype* is "CT_QAWC." dqawc() uses a Fortran routine qawc.f (Netlib, 2008c, 2008) which is in Quadpack library to calculate the integration.

context Algorithm :: dqng(*sfun: String, a: Real, b: Real, epsabs: Real, epsrel: Real*)
pre: *self.InputData.fntype* = Ctype::CT_QNG
post: if *self.InputData.fntype* = Ctype::CT_QNG
then result = Tuple {

$$self.res = \text{approximate value of } y = \int_a^b f(x)\, dx,$$

$$self.abserr = \text{approx}|\, y - res\, |,$$

$$self.neval = \text{number of integrand evaluations},$$

{Ecodetype::NORMAL, Ecodetype::MAX_EVAL_LIMIT} -
> includes *self.ier*
}
else ExceptionID = Fntype_invalid
endif

Assumptions: None

Descriptions: dqng() calculates the integration of the input function whose *Ctype* is "CT_QNG." dqng() uses a Fortran routine qng.f (Netlib, 2008f, 2008) which is in Quadpack library to calculate the integration.

context Algorithm :: dqags(*sfun: String, a: Real, b: Real, epsabs: Real, epsrel: Real*)
pre: *self.InputData.fntype* = Ctype::CT_QAGS
post: if *self.InputData.fntype* = Ctype::CT_QAGS
then result = Tuple {

$$self.res = \text{approximate value of } y = \int_a^b f(x)\, dx,$$

$$self.abserr = \text{approx}|\, y - res\, |,$$

$$self.neval = \text{number of integrand evaluations},$$

{Ecodetype::NORMAL, Ecodetype::MAX_EVAL_LIMIT, Ecodetype::RNDOFF_ERR, Ecodetype::LOC_DIFF, Ecodetype::NOT_CONVG, Ecodetype::DIVG_ING
-> includes *self.ier*
}

else ExceptionID = Fntype_invalid
        endif

Assumptions: None

Descriptions: dqags() calculates the integration of the input function whose *Ctype* is "CT_QAGS." dgags() uses a Fortran routine qng.f (Netlib, 2008b, 2008) which is in Quadpack library to calculate the integration.

context Algorithm :: dqag(*sfun: String, a: Real, b: Real, epsabs: Real, epsrel: Real, key: Integer*)

pre: key = 6

post: if *self.InputData.fntype* = Ctype::CT_QAG

      then if key = 6

           then result = Tuple {

$$self.res = \text{approximate value of } y = \int_a^b f(x)\, dx,$$

                $self.abserr = \text{approx}|\,y - res\,|,$

                $self.neval = \text{number of integrand evaluations,}$

                {Ecodetype::NORMAL, Ecodetype::MAX_EVAL_LIMIT,

Ecodetype::RNDOFF_ERR, Ecodetype::LOC_DIFF} -> includes *self.ier*

                }

           else ExceptionID = Key_Invalid

           endif

      else ExceptionID = Fntype_invalid

      endif

Assumptions: None

Descriptions: dqag() calculates the integration of the input function whose *Ctype* is "CT_QAG." dgag() uses a Fortran routine qag.f (Netlib, 2008a, 2008) which is in Quadpack library to calculate the integration. Notice according to the descriptions of qag.f, the parameter *key* should be set to 6.

context Algorithm :: getRes() : Real

pre: true

post: result = *self.res*

Assumptions: getRes() should be invoked after dqng(), dqags(), dqawc(), dqawo(), dqaws() and dqag()

Description: get the estimated value of the integral *res*

context Algorithm :: getAbserr() : Real

pre: true

post: result = *self.abserr*

Assumptions: getAbserr() should be invoked after dqng(), dqags(), dqawc(), dqawo(), dqaws() and dqag()

Description: get the estimated absolute error *abserr*

context Algorithm :: getNeval() : Real
pre: true
post: result = *self.neval*
Assumptions: getNeval() should be invoked after dqng(), dqags(), dqawc(), dqawo(), dqaws() and dqag()
Description: get *neval*

context Algorithm :: getErrorCode() : Ecodetype
pre: true
post: result = *self.ier*
Assumptions: getErrorCode() should be invoked after dqng(), dqags(), dqawc(), dqawo(), dqaws() and dqag()
Description: get an error code *ier*

C.2.8.7 OutputShow Class

OutputShow Class shows the estimated value of the integral, the estimated absolute error, the number of times the integral was evaluated and an error code on the screen.

Attributes
None.
Class Invariant
None.
Operations

context OutputShow::show(*res: Real, abserr: Real, neval: Integer, ier: Ecodetype*)
pre: true
post: true
Assumptions: show() is the last to execute in ONIS
Description: Show *res*, *abserr*, *neval* and *ier* on the Screen

C.2.8.8 Expression Class

Expression Class is used to express the structure of a parse tree. There are four constructors and two operations in this class. Four constructors are to generate different kinds of nodes in a parser tree. Operation *setvalue* helps to set the different x value to the expressions. *evaluate* is to obtain the values of function evaluations.

Data Types
Two enumeration data types TFLAG and SYMBOL are used in Expression class. TFLAG is a flag that indicates the data type of a node in a parse tree. VAR, CON and EXP means a variable, a constant and an expression, respectively. Notice only one kind of variable x is allowed in ONIS. SYMBOL helps to check the symbols in the input function. Figure

8 illustrates the data type of TFLAG and SYMBOL.

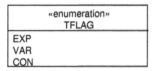

```
        «enumeration»
           TFLAG
  EXP
  VAR
  CON
```

```
        «enumeration»
          SYMBOL
  ADD
  SUB
  MUL
  DIV
  LBRACK
  RBRACK
  COMMA
  NUM
  VAR
  SIN
  COS
  EXP
  LOG
  LOG10
  POW
  SQRT
  TAN
  EOF
  PI
  INVALID
```

Figure C.10: TFLAG and SYMBOL

Attributes
Table C.6 shows the attributes of Expression Class.

Class Invariant
None.

C.2.8.9   Operations

There are four constructors in Expression Class to initialize the attributes so as to generate different type nodes in a parse tree. In this parse tree, the leaf node can only be a constant or a variable x.

Name	Data Type	Descriptions
typ	TFLAG	data type of a node
op	SYMBOL	operator and symbol
opd1	Expression	a left child of a node
opd2	Expression	a right child of a node
value	Real	value of a node

Table C.6: Attributes of Expression Class

context Expression :: Expression()
 context Expression :: typ : TFLAG

init: typ = TFLAG::CON
 context Expression :: value : Real

init: value = 0
 context Expression :: op : SYMBOL

init: op = SYMBOL::INVALID
 context Expression :: opd1 : Expression

init: opd1 = Empty
 context Expression :: opd2 : Expression

init: opd2 = Empty

Assumptions: As a constructor, Expression() should be executed before any other operations in the Expression Class, except for other constructor Expression()
Descriptions: This is a default constructor to create initial values of a leaf node. The initial data type of the node is set to CON, a number, with a value of zero. The op, i.e. operator, is set INVALID, and left child opd1 and right child opd2 of this node are set to empty. In OCL, no standard way are found to express that the value of an object is empty; therefore, here we use Empty to express it. Developers can use different methods to implement Empty in the implementation stage in terms of the programming language they choose.

context Expression :: Expression(number: Real)
 context Expression :: typ : TFLAG

init: typ = TFLAG::CON
 context Expression :: value : Real

init: value = number
 context Expression :: op : SYMBOL

init: op = SYMBOL::INVALID

context Expression :: *opd1* : *Expression*

init: *opd1* = Empty
  context Expression :: *opd2* : *Expression*

init: *opd2* = Empty

Assumptions: As a constructor, Expression(*number: Real*) is executed before any other operations in the Expression Class, except for other constructor Expression()
Descriptions: Expression(*number: Real*) is also a constructor. Its difference with previous Expression() lies in it can help set a leaf node whose data type is also CON but with a value which comes from the parameter of Expression() *number*.

context Expression :: Expression(*s: String*)
  context Expression :: *typ* : *TFLAG*

init: *typ* = TFLAG::VAR
  context Expression :: *value* : *Real*

init: *value* = 0
  context Expression :: *op* : *SYMBOL*

init: *op* = SYMBOL::INVALID
  context Expression :: *opd1* : *Expression*

init: *opd1* = Empty
  context Expression :: *opd2* : *Expression*

init: *opd2* = Empty

Assumptions: As a constructor, Expression(*s: String*) should be executed before any other operations in the Expression Class, except for other constructor Expression(). In ONIS, only one kind of variable, *x*, is allowed.
Descriptions: Expression(*s: String*) is also a constructor. Its difference with Expression() lies in it can help set a leaf node whose data type is VAR, i.e. a variable.

context Expression :: Expression(*symbol*: SYMBOL, *obj1*: Expression, *obj2*: Expression) : Expression
  context Expression :: *typ* : *TFLAG*

init: *typ* = TFLAG::VAR
  context Expression :: *value* : *Real*

init: *value* = 0

context Expression :: *op* : *SYMBOL*

init: *op* = SYMBOL::INVALID
 context Expression :: *opd1* : *Expression*

init: *opd1* -> *obj1*
 context Expression :: *opd2* : *Expression*

init: *opd2* -> *obj2*

Assumptions: As a constructor, Expression(*symbol*: SYMBOL, *obj1*: Expression, *obj2*: Expression) should be executed before any other operations in the Expression Class,except for other constructor Expression()

Descriptions: This constructor is to create a node with left child and right child. In this case, data type of the node will be set to EXP, i.e. an expression. *opd1* and *opd2* will point to left child and right child respectively. *op* store the information of symbol, which could be a function such as SIN or COS, or a operator such as "+" or "-".

context Expression :: setValue(*val: Real*)
pre: true
body: if *self.typ* = TFLAG::VAR then *self.value* = val
               else if *self.typ* = TFLAG::EXP
                  then self.op1.setValue(val);
                  if op2 ->notEmpty() then self.op2.setValue(val) endif
                  endif
             endif
Assumptions: setValue() should be executed before evaluate()
Description: set the value *val* for the points *x* at which the integrand is evaluated to the entire parser tree

context Expression :: evaluate() : Real
pre: true
post: if {TFLAG::VAR, TFLAG::CON} -> includes *self.typ* then result = *self.value*
               else if *self.op* = TFLAG::ADD
                  then if *self.opd2* -> isEmpty()
                      then result = *self.op1*.evaluate()
                  else result = *self.op1*.evaluate() + *self.op2*.evaluate()
                  endif
                  endif
               else if *self.op* = TFLAG::SUB
                  then if *self.opd2* -> isEmpty()
                      then result = *self.op1*.evaluate()
                  else result = *self.op1*.evaluate() - *self.op2*.evaluate()
                  endif
                  endif

else if *self.op* = TFLAG::MUL
  then result = *self.op1*.evaluate() * *self.op2*.evaluate()
  endif
else if *self.op* = TFLAG::DIV
  then if *self.op2*.evaluate() = 0 then Zero_Invalid
    else result = *self.op1*.evaluate() / *self.op2*.evaluate()
    endif
  endif
else if *self.op* = TFLAG::SIN
  then result = sin(*self.op1*.evaluate())
  endif
else if *self.op* = TFLAG::COS
  then result = cos(*self.op1*.evaluate())
  endif
else if *self.op* = TFLAG::EXP
  then result = exp(*self.op1*.evaluate())
  endif
else if *self.op* = TFLAG::LOG
  then result = log(*self.op1*.evaluate())
  endif
else if *self.op* = TFLAG::LOG10
  then result = log10(*self.op1*.evaluate())
  endif
else if *self.op* = TFLAG::SQRT
  then result = sqrt(*self.op1*.evaluate())
  endif
else if *self.op* = TFLAG::TAN
  then result = tan(*self.op1*.evaluate())
  endif
else if *self.op* = TFLAG::POW
  then result = pow(*self.op1*.evaluate(), *self.op2*.evaluate())
  endif
else Input_Function_Invalid
endif

Assumptions: evaluate() should be executed after setValue()
Description: calculate the values of the integrand in the evaluated point x

Examples The following figures illustrate how to use Expression Class to express the nodes of the parse tree. If the input function is $x + sin(cos(x + 3))$, Figure C.11 shows the parse tree of this function and Figure C.12 presents a parse tree using Expression Class to express this parse tree.

C.2.8.10   Parser Class

Parser Class is used to parse the input function and create a parse tree. The following is the Extended Backus-Naur Form (EBNF) grammar for the input function.

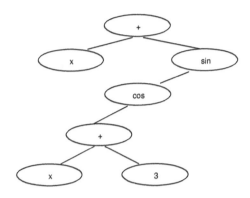

Figure C.11: Parse Tree for $x + \sin(\cos(x+3))$

Name	Data Type	Descriptions
sym	*SYMBOL*	symbol of the node
size	*Integer*	length of the string of input function
pos1	*Integer*	a pointer to record the current node
pos2	*Integer*	a pointer to scan the input function to find the symbol of a node
sfunction	*String*	input function
strval	*String*	part of the input function
numval	*Real*	number in the input function

Table C.7: Attributes of Parser Class

expression = ["+" | "-"] term {("+" | "-") term}
term = factor {("*" | "/") factor}
factor = number | variable | funct | "(" expression ")"
funct = ("COS" | "SIN" | "TAN" | "EXP" | "LOG" | "LOG10" | "POW" | "SQRT")
"(" expression ")"

Attributes
Table C.7 shows the attributes of Parser Class. The following attributes are public, so all the operations in Parser Class can share the value of the attributes.

Attribute Initialization
context Parser :: *pos1* : *Integer*

init: *pos1* = 0
context Parser :: *pos2* : *Integer*

190

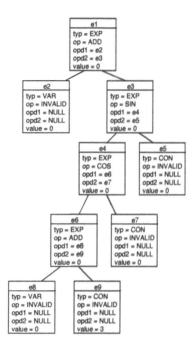

Figure C.12: Parse Tree for $x + \sin(\cos(x+3))$ using Expression Class

init: $pos2 = 0$

Class Invariant
None.
Operations

context Parser :: parse(*strFunction: String*) : Expression
pre: true
post: result = self.expression()
Assumptions: parse() is the function that should be invoked before any other operations in Parser Class.
Description: parse() initializes the value of *pos1* and *pos2*. Then, it calls expression() to parse the input function and generate a parse tree.

context Parser :: getSymbol()
pre: true
post: if sfunction -> substring(pos1, pos1) = "("
                              then pos1 = pos1@pre + 1; pos2 = pos2@pre + 1; sym = SYMBOL::LBRACK

191

else if sfunction -> substring(pos1, pos1) = ","
      then pos1 = pos1@pre + 1; pos2 = pos2@pre + 1; sym =
SYMBOL::COMMA endif
      else if sfunction -> substring(pos1, pos1) = ")"
      then pos1 = pos1@pre + 1; pos2 = pos2@pre + 1; sym =
SYMBOL::RBRACK endif
      else if sfunction -> substring(pos1, pos1) = "+"
      then pos1 = pos1@pre + 1; pos2 = pos2@pre + 1; sym =
SYMBOL::ADD endif
      else if sfunction -> substring(pos1, pos1) = "-"
      then pos1 = pos1@pre + 1; pos2 = pos2@pre + 1; sym =
SYMBOL::SUB endif
      else if sfunction -> substring(pos1, pos1) = "*"
      then pos1 = pos1@pre + 1; pos2 = pos2@pre + 1; sym =
SYMBOL::MUL endif
      else if sfunction -> substring(pos1, pos1) = "/"
      then pos1 = pos1@pre + 1; pos2 = pos2@pre + 1; sym =
SYMBOL::DIV endif
      else if sfunction -> substring(pos1, pos1 + 2) = "sin"
      then pos1 = pos1@pre + 3; pos2 = pos2@pre + 3; sym =
SYMBOL::SIN endif
      else if sfunction -> substring(pos1, pos1 + 2) = "cos"
      then pos1 = pos1@pre + 3; pos2 = pos2@pre + 3; sym =
SYMBOL::COS endif
      else if sfunction -> substring(pos1, pos1 + 2) = "exp"
      then pos1 = pos1@pre + 3; pos2 = pos2@pre + 3; sym =
SYMBOL::EXP endif
      else if sfunction -> substring(pos1, pos1 + 2) = "log"
      then pos1 = pos1@pre + 3; pos2 = pos2@pre + 1; sym =
SYMBOL::LOG endif
      else if sfunction -> substring(pos1, pos1 + 4) = "log10"
      then pos1 = pos1@pre + 5; pos2 = pos2@pre + 5; sym =
SYMBOL::LOG10 endif
      else if sfunction -> substring(pos1, pos1 + 2) = "pow"
      then pos1 = pos1@pre + 3; pos2 = pos2@pre + 3; sym =
SYMBOL::POW endif
      else if sfunction -> substring(pos1, pos1 + 3) = "sqrt"
      then pos1 = pos1@pre + 4; pos2 = pos2@pre + 4; sym =
SYMBOL::SQRT endif
      else if sfunction -> substring(pos1, pos1 + 2) = "tan"
      then pos1 = pos1@pre + 3; pos2 = pos2@pre + 3; sym =
SYMBOL::TAN endif
      else if sfunction -> substring(pos1, pos1) = "x"
      then pos1 = pos1@pre + 1; pos2 = pos2@pre + 1; sym =
SYMBOL::VAR endif
      else if sfunction -> substring(pos1, pos1+1) = "pi"
      then pos1 = pos1@pre + 2; pos2 = pos2@pre + 2; sym =
SYMBOL::PI endif

else ExceptionID = Symbol_Invalid
    endif

Assumptions: None.

Description: getSymbol() is to get the symbol of a node. The position of the node will be obtained from the value of the attribute *pos1* and *pos2*. When it gets the symbol of the node, it will store the symbol to the attribute *sym*.

context Parser :: expression() : Expression

pre: true

post: if sym = SYMBOL::ADD or sym = SYMBOL::SUB
     then *result = self.term()*
     endif

Assumptions: expression() should be executed before term(), factor() and funct()

Description: generate expressions of the parser tree

context Parser :: term() : Expression

pre: true

post: if sym = SYMBOL::MUL or sym = SYMBOL::DIV   then *result = self.factor()*   endif

Assumptions: term() should be executed after expression() but before factor() and funct()

Description: generate terms of the parser tree

context Parser :: factor() : Expression

pre: true

post: if sym = SYMBOL::PI then result = Expression::Expression(PI)
     else if sym = SYMBOL::NUM then result = Expression::Expression(numval)
     else if sym = SYMBOL::VAR then result = Expression::Expression(strval)
     else if sym = SYMBOL::LBRACK then result = expression()
     else if SYMBOL::SIN, SYMBOL::COS, SYMBOL::EXP, SYMBOL::LOG,
SYMBOL::LOG10, SYMBOL::POW, SYMBOL::SQRT, SYMBOL::TANthen result =
funct()
     else Input_Invalid
     endif

Assumptions: factor() should be executed after expression() and term() but before funct()

Description: generate factor of the parser tree

context Parser :: funct() : Expression

pre: true

post: result = expression()

Assumptions: funct() should be executed after expression(), term() and factor()

Description: generate function of the parser tree

C.2.9 Dynamic Modeling of Behaviour

The previous class diagram describes the static aspects of ONIS. Static models, such as class diagrams, describe the objects in a system, the data each object contains and the links that exist between them, but they say very little about the behaviour of these object. When a system is running, objects interact by passing messages. The messages that are sent determine the system's behaviour, but they are not shown on static diagrams such as class diagrams. In this section, sequence diagrams and statecharts are introduced to present the behaviour of ONIS.

C.2.10 Sequence Diagram of ONIS

The sequence diagram is used in this document to show interactions between objects. Figure 11 illustrates the main sequence of ONIS. The classifier roles involved in the interaction are displayed at the top of the diagram. The vertical dimension in a sequence diagram represents time and the messages in an interaction are drawn from top to bottom of the diagram, in the order that they are sent. Each role has a dashed line, known as its *lifeline* extending below it. The lifeline indicates the period of time during which objects playing that role actually exist.

Messages are shown as arrows leading from the lifeline of the sender of the message to that of the receiver. When a message is sent, control passes from the sender of the message to the receiver. The period of time during which an object is processing a message is known as an *activation* and is shown on a lifeline as a narrow rectangle whose top is connected to a message.

When an object finishes processing a message, control returns to the sender of the message. This marks the end of the activation corresponding to that message and is marked by a dashed arrow going from the bottom of the activation rectangle back to the lifeline of the role that sent the message giving rise to the activation.

The messages shown in Figure 11, with a solid arrowhead, denote synchronous message, such as normal procedure calls. These are characterized by the fact that processing in the object that sends the message is suspended until the called object finishes dealing with the message and returns control to the caller. (Priestley, 2003, page 192,193)

It is optional whether or not to show activations return messages on sequence diagrams. In Figure 11, to simplify the sequence diagram and make the diagram clearer, some return messages are not included in the diagram.

Sequence diagrams also provide means for showing conditional message passing or, in other words, messages that are only sent under certain circumstance. In Figure 11, to show that message dqawo() will only be sent under certain circumstances, a condition, C1, is attached to it. This consists of a Boolean expression written in square bracket. If the condition evaluates to true at the point in the activation when the message is reached, the message will be sent. Otherwise, control will jump to the point following message corresponding to the message bearing the condition (Priestley, 2003, page 200). For example dqawo() only be sent to Algorithm when function type equal to CT_QAWO.

The following is the meanings for condition C1 to C6.

- C1: fntype = Ctype::CT_QAWO

- C2: fntype = Ctype::CT_QAWS

- C3: fntype = Ctype::CT_QAWC

- C4: fntype = Ctype::CT_QAGS

- C5: fntype = Ctype::CT_QNG

- C6: fntype = Ctype::CT_QAG

The are some messages in the Figure 11 with '*' beside the messages, that means these messages will also be done after the message dqaws(), dqawc(), dqags(), dqng() and dqag(). To simplify the sequence diagram, these messages are only be drawn once in the diagram.

C.2.10.1 Statechart of Expression

The Parser Class is to parse the input function and generate parse tree. The parse tree is expressed using Expression Class. A parser can be regarded as a finite state machine, which consists of a set of states and set of labeled transitions between states. An input function can be treated as a sequence which is recognized by starting in the initial state and from each state following the transition which is labeled with the next input symbol (Sekerinski, 2006, page 45). In UML, state machines are normally documented in a type of diagram known as statechart. Figure 12 is the statechart for Expression objects which is used in ONIS.

Statecharts show the possible states of an object, the events it can detect and its response to those events. In software terms, it is common to assume that the events detected by an object are simply the message sent to it. In general, detecting an event can cause an object to move from one state to another. Such a move is called a *transition*. The basic information shown on a statechart is the possible states of the entity and the transitions between them, or in other words the way that detecting various events causes the system to move from one state to another. The states of the system are shown are rounded rectangles, with the name of the state written inside them. State transitions are shown by arrows linking two states. Each such arrow must be labeled with the name of an event. The meaning of such an arrow is that if the system receives the event when it is in the state at the tail of the arrow, it will move into the state at the head of the arrow (Priestley, 2003, page 210).

An initial states are shown as small black disks. A transition leading from an initial event shows the state that the object goes into when it is created or initialized. No event should be written on a transition from an initial state.

According to EBNF grammar in Section 8.7, an *expression* could be a *term*, a *factor* or a *funct* (i.e. a function). So, *term*, *factor* and *funct* are states of *expression* but they are substates. In addition, *term* is a substate of *expression, factor* is a substate of *term* and *funct* is a substate of *factor*. In statecharts, we can use *composite state* consisting of one or many nested substates to express this kind of relationship.

We can add *guard conditions* to the transitions, stating the circumstances under which the transitions will fire. Guard conditions are part of the specification of a transition and are written in square brackets after the event name that labels the transition. Guard conditions are often written in informal English, but if desired a more formal notation can be used, such as the OCL language describe above (Priestley, 2003, page 212-214). For example, in Figure 12, Expression is in the state *funct* and transition getsymbol() occurs and sym equals to LBRACK (a left bracket), the state of Expression moves from *funct* to *expression*. We use getsymbol()[sym = SYMBOL::LBRACK] to describe this transition.

History states are represented by a capital 'H' within a circle and can only appear inside composite states. A transition to a history state causes the substate that was most recently activate in the composite state to become active again i.e. History state could 'remember' which substate was activate last time the composite state was active and automatically return to that substate. Another variant of a history state "deep history state', notated with an additional * inside the history pseudostate. This means 'recursively enter the most recently vacated substate of every non-concurrent composite substate of the composite state enclosing myself" (Lano, 2005, page 80), is used. In Figure 12, we use deep history state to remember the substates.

C.2.11   Exception Handling

This section describes the exception handling strategies using in ONIS. Generally, any occurrence of an abnormal condition that causes an interruption in normal control flow is called an exception. It is said that an exception is raised (thrown) when such a condition is signaled by a software unit. In response to an exception, the control is immediately given to a designated handler for the exception, which reacts to that situation (exception handler). The handler can try to recover from that exception in order to continue at a predefined location or it cleans up the environment and further escalates the exception (Renzel, 2008, page 9).

In ONIS, the exception handling in a region of code by surrounding it with a *try* block. The exception handlers immediately after the *try* block, in a series of *catch* clauses. An exception handler is much like a definition for a function, named by the keyword *catch*. The statement, for example:
   throw LowerBound_input_not_valid

causes the complier to search the function calls for a handler that can catch exception

code LowerBound_input_not_valid, so control passes to that handler. When executing a throw expression, the program will jump to a handler associated with an active try block. In ONIS, exception handler will catch the exception codes and responses with a message which describes the type of exceptions on the screen.

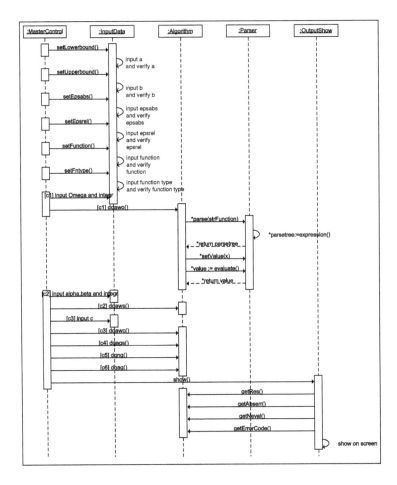

Figure C.13: Main Sequence of ONIS

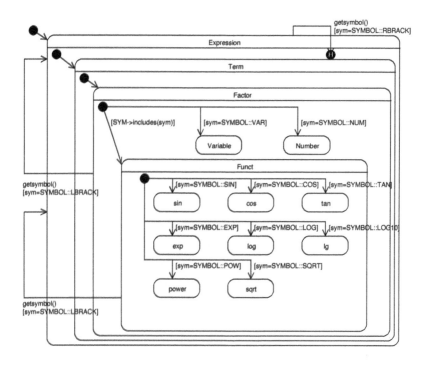

Figure C.14: Statecharts of Expression